Origins of Life

How did life on Earth originate? Did replication or metabolism come first in the history of life? In this extensively rewritten second edition, Freeman Dyson examines these questions and discusses the two main theories that try to explain how naturally occurring chemicals could organize themselves into living creatures.

The majority view is that life began with replicating molecules, the precursors of modern genes. The minority belief is that random populations of molecules evolved metabolic activities before exact replication existed and that natural selection drove the evolution of cells toward greater complexity for a long time without the benefit of genes. Dyson analyzes both of these theories with reference to recent important discoveries by geologists and biologists, aiming to stimulate new experiments that could help decide which theory is correct.

Since the first edition of this book was published in 1985, revolutionary discoveries have been made in biology, genetics, and geology, casting new light on the questions of the origins of life. Molecular biologists discovered ribozymes, enzymes made of RNA. Geneticists discovered that many of the most ancient creatures are thermophilic, living in hot environments. Geologists discovered evidence of life in the most ancient of all terrestrial rocks in Greenland.

This second edition covers the enormous advances that have been made in biology and geology in the past decade and a half and the impact they have had on our ideas about how life began. Freeman Dyson's clearly written, fascinating book will appeal to anyone interested in the origins of life.

Freeman Dyson, currently Emeritus Professor at the Institute for Advanced Study in Princeton, is a distinguished scientist and a gifted writer. He is a fellow of the Royal Society of London and a member of the U.S. National Academy of Sciences, as well as the holder of eighteen honorary degrees. His most recent books include *Imagined Worlds* (1997) and *From Eros to Gaia* (1992).

Origins of Life

Revised Edition

FREEMAN DYSON

*Institute for Advanced
Study, Princeton*

CAMBRIDGE
UNIVERSITY PRESS

PUBLISHED BY THE PRESS SYNDICATE OF THE UNIVERSITY OF CAMBRIDGE
The Pitt Building, Trumpington Street, Cambridge, United Kingdom

CAMBRIDGE UNIVERSITY PRESS
The Edinburgh Building, Cambridge CB2 2RU, UK   http://www.cup.cam.ac.uk
40 West 20th Street, New York, NY 10011-4211, USA   http://www.cup.org
10 Stamford Road, Oakleigh, Melbourne 3166, Australia

First published 1999

Printed in the United States of America

Typeset in Meridien 9.75/13pt. in LaTeX 2_ε [TB]

A catalog record for this book is available from the British Library.

Library of Congress Cataloging-in-Publication Data
Dyson, Freeman J.
    Origins of life / Freeman Dyson.
        p.   cm.
    Includes bibliographical references and index.
    ISBN 0-521-62668-4 (pbk.)
    1. Life – Origin.   I. Title.
    QH325.D88   1999
    576.8′3 – dc21   99-21079
                                                          CIP
ISBN 0 521 62668 4 paperback

Contents

Preface

The delivery of my Tarner Lectures in Cambridge happened to coincide exactly with the two-hundredth anniversary of the first manned flight across the English Channel by Blanchard and Jeffries in January 1785. Like the intrepid balloonists, a public lecturer must carry supplies of hot air and of ballast to regulate his flight – hot air to be inserted when the text of the lecture is too short and ballast to be dropped when the text is too long. In preparing the lectures for publication I was able to retrieve some of the dropped ballast and to vent some of the inserted air. I am grateful to my hosts at Trinity College for their hospitality and to my audiences for their sharp questions and criticisms. In revising the book for this second edition in 1998, I have had the benefit of many additional criticisms from readers of the first edition. I am grateful to everyone who corrected my mistakes and told me about recent developments in evolutionary biology. I am especially grateful to Professor Cairns-Smith for reading and criticizing the new edition. The first edition was a lightly edited transcript of the lectures. The second edition is substantially enlarged and is no longer a transcript. Much has happened in the last thirteen years to deepen our understanding of early evolution. I have changed my story to take account of new discoveries. But the basic mystery of life's origin remains unsolved, and the central theme of the book remains unchanged.

The Tarner Lectures were established with the requirement that the lecturer speak "on the philosophy of the sciences and the relations or want of relations between the different departments of knowledge." I intended to ignore this requirement when I planned the lectures. I preferred to deal with concrete scientific problems rather than with philosophical generalities. I chose the origins of life as my theme because I judged the time to be ripe for a new experimental attack on the problem of origins. The main purpose of the lectures was to stimulate experiments. Nevertheless, it happens that the study of the origins of life touches many scientific disciplines and raises many philosophical questions. I therefore found myself, in spite of my pragmatic and unphilosophical intentions, fortuitously following Mr. Tarner's wishes. It was impossible to speak for four hours about the origins of life without encountering some ideas that connect widely separated branches of science and other ideas that stray over the border from science to philosophy.

The lectures were addressed to a general university audience. The readers of this book are likewise expected to be educated but not expert. The same thing can be said of the author. I do not pretend to be an expert in biology. I have not read systematically through the technical literature. In my survey of experiments and ideas, I made no attempt to be complete or even to be fair. I apologize in advance to all the people, living and dead, whose contributions to knowledge I have ignored, especially to J. B. S. Haldane, Desmond Bernal, Sidney Fox, Hyman Hartman, Pier Luisi, Julian Hiscox, Lee Smolin, and Stuart Kauffman. I apologize also to Paul Davies, whose excellent book (Davies, 1998) was published just as mine was going to press. I missed the chance to engage in a friendly debate with Davies, explaining where we agree and where we disagree.

I am grateful to Martin Rees and Sydney Brenner for inviting me to a meeting with the title "From Matter to Life," which was held at King's College, Cambridge, in September 1981. Biologists, chemists, physicists, and mathematicians came together to talk about the origins of life, and in three days I acquired the greater part of my education as an evolutionary biologist. That meeting led me to the point of view I am expressing in this book. I wish also to thank the Master and Fellows of Trinity College for inviting me to Cambridge as Tarner Lecturer in 1985.

The first two chapters of the book are historical. Chapter 1 introduces the six characters who contributed the most to my thinking about the origins of life. Chapter 2 describes in greater detail the leading theories and the experimental background from which they arose. Chapter 3 is the most technical chapter. It describes my own contribution to the subject, a mathematical model that is intended to represent in abstract form the transition from chaos to organized metabolic activity in a population of molecules. Chapter 4 discusses some of the questions the model leaves open and the implications of the model for the later stages of biological evolution. At the end of Chapter 4, I included, in deference to Mr. Tarner, an excursion into philosophy. My approach to the understanding of the origins of life emphasizes diversity and error tolerance as life's salient characteristics. This approach led me to draw analogies between the phenomena of cellular biology and the phenomena of ecology and cultural evolution, but the validity of these speculative analogies is in no way essential to our understanding of cellular biology.

Freeman J. Dyson
Institute for Advanced Study, Princeton, New Jersey, USA
November 1998

CHAPTER ONE

Illustrious Predecessors

SCHRÖDINGER AND VON NEUMANN

In February 1943, at a bleak moment in the history of mankind, the physicist Erwin Schrödinger gave a course of lectures to a mixed audience at Trinity College, Dublin. Ireland was then, as it had been in the days of Saint Columba fourteen hundred years earlier, a refuge for scholars and a nucleus of civilization beyond the reach of invading barbarians. It was one of the few places in Europe where peaceful scientific meditation was still possible. Schrödinger proudly remarks in the published version of the lectures that they were given "to an audience of about four hundred which did not substantially dwindle." The lectures were published by the Cambridge University Press in 1944 in a little book (Schrödinger, 1944) with the title *What is Life?*

Schrödinger's book is less than a hundred pages long. It was widely read and was influential in guiding the thoughts of the young people who created the new science of molecular biology in the following decade. It is clearly and simply written, with only five references to the technical literature and less than ten equations from beginning to end. It is, incidentally, a fine piece of English prose. Although Schrödinger was exiled from his native Austria to Ireland when he was over fifty, he wrote English far more beautifully than most of his English and American contemporaries. He reveals his cosmopolitan background only in the epigraphs that introduce his chapters: three are from Goethe, in German; three are from Descartes and Spinoza, in Latin; and one is from Unamuno,

1

in Spanish. As a sample of his style I quote the opening sentences of his preface:

> A scientist is supposed to have a complete and thorough knowledge, at first hand, of some subjects, and therefore he is usually expected not to write on any topic of which he is not a master. This is regarded as a matter of noblesse oblige. For the present purpose I beg to renounce the noblesse, if any, and to be freed of the ensuing obligation. My excuse is as follows. We have inherited from our forefathers the keen longing for unified, all-embracing knowledge. The very name given to the highest institutions of learning reminds us that from antiquity and throughout many centuries the universal aspect has been the only one to be given full credit. But the spread, both in width and depth, of the multifarious branches of knowledge during the last hundred odd years has confronted us with a queer dilemma. We feel clearly that we are only now beginning to acquire reliable material for welding together the sum-total of what is known into a whole; but, on the other hand, it has become next to impossible for a single mind fully to command more than a small specialized portion of it. I can see no other escape from this dilemma (lest our true aim be lost for ever) than that some of us should venture to embark on a synthesis of facts and theories, albeit with second-hand and incomplete knowledge of some of them, and at the risk of making fools of themselves. So much for my apology.

This apology for a physicist venturing into biology will serve for me as well as for Schrödinger, although in my case the risk of the physicist making a fool of himself may be somewhat greater.

Schrödinger's book was seminal because he knew how to ask the right questions. What is the physical structure of the molecules that are duplicated when chromosomes divide? How is the process of duplication to be understood? How do these molecules retain their individuality from generation to generation? How do they succeed in controlling the metabolism of cells? How do they create the organization that is visible in the structure and function of higher organisms? He did not answer these questions, but by asking them he set biology moving along the path that led to the epoch-making discoveries of the subsequent forty years: to the discovery of the double helix and the triplet code, to the precise analysis and whole-sale synthesis of genes, and to the quantitative measurement of the evolutionary divergence of species.

One of the great pioneers of molecular biology who was active in 1943 and is still active today, Max Perutz, dissents sharply from my appraisal of Schrödinger's book (Perutz, 1989). "Sadly," Perutz writes, "a close study of his book and of the related literature has shown me that what was true in his book was not original, and most of what was original was known not to be true even when the book was written." Perutz's statement is well founded. Schrödinger's account of existing knowledge is borrowed from his friend Max Delbrück, and his conjectured answers to the questions that he raised were indeed mostly wrong. Schrödinger was woe-fully ignorant of chemistry, and in his isolated situation in Ireland he knew little about the new world of bacteriophage genetics that Delbrück had explored after emigrating to the United States in 1937. But Schrödinger never claimed that his ideas were original, and the importance of his book lies in the questions that he raised rather than in the answers that he conjectured. In spite of Perutz's dissent, Schrödinger's book remains a classic because it asked the right questions.

Schrödinger showed wisdom not only in the questions that he asked but also in the questions that he did not ask. He did not ask any questions about the origin of life. He understood that the time was ripe in 1943 for a fundamental understanding of the physical basis of life. He also understood that the time was not then ripe for any fundamental understanding of life's origin. Until the basic chemistry of living processes was clarified, one could not ask meaningful questions about the possibility of spontaneous generation of these processes in a prebiotic environment. He wisely left the question of origins to a later generation.

Now, half a century later, the time is ripe to ask the questions Schrödinger avoided. We can hope to ask the right questions about origins today because our thoughts are guided by the experimental discoveries of Manfred Eigen, Leslie Orgel, and Thomas Cech. The questions of origin are now becoming experimentally accessible just as the questions of structure were becoming experimentally accessible in the 1940s. Schrödinger asked the right questions about structure because his thoughts were based on the experimental discoveries of Timoféeff-Ressovsky, who exposed fruit-flies to X-rays and measured the relationship between the dose of radiation and the rate of appearance of genetic mutations. Delbrück was a friend of Timoféeff-Ressovsky and published a joint paper with him describing and interpreting the experiments (Timoféeff-Ressovsky et al., 1935). Their joint paper provided the experimental basis for Schrödinger's questions. After 1937, when Delbrück came to America, he continued to explore the problems of structure. Delbrück hit on the bacteriophage as the ideal experimental tool, a biological system stripped of inessential complications and reduced to an almost bare genetic apparatus. The bacteriophage was for biology what the hydrogen atom was for physics. In a similar way Eigen became the chief explorer of the problems of the origin of life in the 1970s because he hit on ribonucleic acid (RNA) as the ideal experimental tool for studies of molecular evolution in the test-tube. Eigen's RNA experiments have carried Delbrück's bacteriophage experiments one step further: Eigen stripped the genetic apparatus completely naked, thereby enabling us to study its replication unencumbered by the baggage of structural molecules that even so rudimentary a creature as a bacteriophage carries with it.

Before discussing the experiments of Eigen, Orgel, and Cech in detail, I want to finish my argument with Schrödinger. At the risk, again, of making a fool of myself, I shall venture to say that in his discussion of the nature of life Schrödinger missed an essential point. And I feel that the same point was also missed by Manfred Eigen in his discussion of the origin of life. I hasten to add that in disagreeing with Schrödinger and Eigen I am not disputing the greatness of their contributions to biology. I am saying only that they did not ask all of the important questions.

In Schrödinger's book we find four chapters describing in lucid detail the phenomenon of biological replication and a single chapter describing less lucidly the phenomenon of metabolism. Schrödinger finds a conceptual basis in physics both for exact replication and for metabolism. Replication is explained by the quantum mechanical stability of molecular structures, whereas metabolism is explained by the ability of a living cell to extract negative entropy from its surroundings in accordance with the laws of thermodynamics. Schrödinger was evidently more interested in replication than in metabolism. There are two obvious reasons for his bias. First, he was, after all, one of the inventors of quantum mechanics, and it was natural for him to be primarily concerned with the biological implications of his own brainchild. Second, his thinking was based on Timoféeff-Ressovsky's experiments, and these were biased in the same direction. The experiments measured the effects of X-rays on replication and did not attempt to observe effects on metabolism. Delbrück carried the same bias with him when he came to America. Delbrück's new experimental system, the bacteriophage, is a purely parasitic creature in which the metabolic function has been lost and only the replicative function survives. It was indeed precisely this concentration of attention upon a rudimentary and highly specialized form of life that enabled Delbrück to do experiments exploring the physical basis of biological replication. It was necessary to find a creature without metabolism to isolate experimentally the phenomena of replication. Delbrück penetrated more deeply than his contemporaries into the mechanics of replication because he was not distracted by the problems of metabolism. Schrödinger saw the world of biology through Delbrück's eyes. It is not surprising that Schrödinger's view of what constitutes a living organism resembles a bacteriophage more than it resembles a bacterium or a human being. His single chapter devoted to the metabolic aspect of life appears to be an afterthought put in for the sake of completeness but not affecting the main line of his argument.

The main line of Schrödinger's argument, which led from the facts of biological replication to the quantum mechanical structure of the gene, was brilliantly clear and fruitful. It set the style for the subsequent development of molecular biology. Neither Schrödinger himself nor the biologists who followed his lead appear to have been disturbed by the logical gap between his main argument and his discussion of metabolism. Looking back on his 1943 lectures now with the benefit of half a century of hindsight, we may wonder why he did not ask some fundamental questions that the gap might have suggested to him: Is life one thing or two things? Is there a logical connection between metabolism and replication? Can we imagine metabolic life without replication, or replicative life without metabolism? These questions were not asked because Schrödinger and his successors took it for granted that the replicative aspect of life is primary and the metabolic aspect secondary. As their understanding of replication became more and more triumphantly complete, their lack of understanding of metabolism was pushed into the background. In popular accounts of molecular biology as it is now taught to school children, life and replication have become practically synonymous. In modern discussions of the origin of life it is often taken for granted that the origin of life is the same thing as the origin of replication. Manfred Eigen's view is an extreme example of this tendency. Eigen chose RNA as the working material for his experiments because he wished to study replication but was not interested in metabolism. Eigen's theories about the origin of life are in fact theories about the origin of replication.

It is important here to make a sharp distinction between replication and reproduction. I am suggesting as a hypothesis that the earliest living creatures were able to reproduce but not to replicate. What does this mean? For a cell, to reproduce means simply to divide into two cells with the daughter cells inheriting approximately equal shares of the cellular constituents. For a molecule, to replicate means to construct a precise copy of itself by a specific chemical process. Cells can reproduce, but only molecules can replicate. In modern times, reproduction of cells is always accompanied by replication of molecules, but this need not always have been so in the past.

It is also important to say clearly what we mean when we speak of metabolism. One of my American friends, a professional molecular biologist, told me that it would never occur to him to ask the question whether metabolism might have begun before replication. For him the word metabolism means chemical processes directed by the genetic apparatus of nucleic acids. If the word has this meaning, then by definition metabolism could not have existed without a genetic apparatus to direct it. He said he was astonished when one of his German colleagues remarked that metabolism might have come first. He asked the German how he could entertain such an illogical idea. For the German, there was nothing illogical in the idea of metabolism coming before replication, because the German word for metabolism is *Stoffwechsel,* which translates into English as "stuffchange." It means any chemical process occurring in cells, whether directed by a genetic apparatus or not. My friend tells me that students who learn molecular biology in American universities always use the word metabolism to mean genetically directed processes. That is one reason they take it for granted that replication must come first. I therefore emphasize that in this book I am following the German and not the American usage. I mean by metabolism what the Germans mean by *Stoffwechsel* with no restriction to genetically directed processes.

Only five years after Schrödinger gave his lectures in Dublin, the logical relations between replication and metabolism were clarified by the mathematician John von Neumann (von Neumann, 1948). Von Neumann described an analogy between the functioning of living organisms and the functioning of mechanical automata. His automata were an outgrowth of his thinking about electronic computers. A von Neumann automaton had two essential components; later on, when his ideas were taken over by the computer industry, these were given the names hardware and software. Hardware processes information; software embodies information. These two components have their exact analogues in living cells; hardware is mainly protein and software is mainly nucleic acid. Protein is the essential component for metabolism. Nucleic acid is the essential component for replication. Von Neumann described precisely, in abstract terms, the logical connections between the components. For a complete self-reproducing automaton, both components are essential. Yet there is an important sense in which hardware comes logically prior to software. An automaton composed of hardware without software can exist and maintain its own metabolism. It can live independently for as long as it finds food to eat or numbers to crunch. An automaton composed of software without hardware must be an obligatory parasite. It can function only in a world already containing other automata whose hardware it can borrow. It can replicate itself only if it succeeds in finding a cooperative host automaton, just as a bacteriophage can replicate only if it succeeds in finding a cooperative bacterium.

In all modern forms of life, hardware functions are mainly performed by proteins and software functions by nucleic acids. But there are important exceptions to this rule. Although proteins serve only as hardware, and one kind of nucleic acid, namely deoxyribonucleic (DNA), serves mainly as software, the other kind of nucleic acid, namely RNA, occupies an intermediate position. RNA is both hardware and software. RNA occurs in modern organisms in four different forms with different functions. There is genomic RNA, constituting the entire genetic endowment of many viruses – in particular the AIDS virus. Genomic RNA is unambiguously software. There is ribosomal RNA, an essential structural component of the ribosomes that manufacture proteins. There is transfer RNA, an essential part of the machinery that brings amino acids to ribosomes to be incorporated into proteins. Ribosomal RNA and transfer RNA are unambiguously hardware. Finally, there is messenger RNA, the molecule that conveys the genetic instructions from DNA to the ribosome. It was believed until recently that messenger RNA was unambiguously software, but Thomas Cech discovered in 1982 that messenger RNA also has hardware functions (Cech, 1993). Cech observed messenger RNA molecules that he called ribozymes performing the functions of enzymes. Ribozymes catalyze the splitting and splicing of other RNA molecules. They also catalyze their own splitting and splicing, in which case they are acting as hardware and software simultaneously. RNA is a flexible and versatile molecule with many important hardware functions in addition to its primary software function. Nevertheless it remains true that the overwhelming majority of metabolic functions of modern organisms belong to proteins, and the overwhelming majority of replicative functions belong to nucleic acids.

Let me summarize the drift of my argument up to this point. Our illustrious predecessor Erwin Schrödinger gave his book the title *What is Life?* but neglected to ask whether the two basic functions of life, metabolism and replication, are separable or inseparable. Our illustrious predecessor John von Neumann, using the computer as a metaphor, raised the question that Schrödinger had missed and gave it a provisional answer. Von Neumann observed that metabolism and replication, however intricately they may be linked in the biological world as it now exists, are logically separable. It is logically possible to postulate organisms that are composed of pure hardware and capable of metabolism but incapable of replication. It is also possible to postulate organisms that are composed of pure software and capable of replication but incapable of metabolism. And if the functions of life are separated in this fashion, it is to be expected that the latter type of organism will become an obligatory parasite upon the former. This logical analysis of the functions of life helps to explain and to correct the bias toward replication that is evident in Schrödinger's thinking and in the whole history of molecular biology. Organisms specializing in replication tend to be parasites, and molecular biologists prefer parasites for experimental study because parasites are structurally simpler than their hosts and better suited to quantitative manipulation. In the balance of nature there must be an opposite bias. Hosts must exist before there can be parasites. The survival of hosts is a precondition for the survival of parasites. Somebody must eat and grow to provide a home for those who only reproduce. In the world of microbiology, as in the world of human society and economics, we cannot all be parasites.

When we begin to think about the origins of life we meet again the question that Schrödinger did not ask, What do we mean by life? And we meet again von Neumann's answer, that life is not one thing but two, metabolism and replication, and that the two things are logically separable. There are accordingly two logical possibilities for life's origins. Either life began only once, with the functions of replication and metabolism already present in rudimentary form and linked together from the beginning, or life began twice, with two separate kinds of creatures, one kind capable of metabolism without exact replication and the other kind capable of replication without metabolism. If life began twice, the first beginning must have been with molecules resembling proteins, and the second beginning with molecules resembling nucleic acids. The first protein creatures might have existed independently for a long time, eating and growing and gradually evolving a more and more efficient metabolic apparatus. The nucleic acid creatures must have been

obligatory parasites from the start, preying upon the protein creatures and using the products of protein metabolism to achieve their
own replication.

The main theme of this book will be a critical examination of
the second possibility, the possibility that life began twice. I call this
possibility the double-origin hypothesis. It is a hypothesis, not a
theory. A theory of the origin of life should describe in some detail
a postulated sequence of events. The hypothesis of dual origin is
compatible with many theories. It may be useful to examine the
consequences of the hypothesis without committing ourselves to
any particular theory.

I do not claim that the double-origin hypothesis is true, or that it
is supported by any experimental evidence. Indeed my purpose is
just the opposite. I would like to stimulate experimental chemists
and biologists and paleontologists to find the evidence by which the
hypothesis might be tested. If it can be tested and proved wrong,
it will have served its purpose. We will then have a firmer foun-
dation of fact on which to build theories of single origin. If the
double-origin hypothesis can be tested and not proved wrong, we
can proceed with greater confidence to build theories of double origin. The hypothesis is useful only insofar as it may suggest new
experiments.

Lacking new experiments, we have no justification for believing
strongly in either the single-origin or the double-origin hypothesis.
I have to confess my own bias in favor of double-origin. But my
bias is based only on general philosophical preconceptions, and I
am well aware that the history of science is strewn with the corpses
of dead theories that were in their time supported by the prevailing philosophical viewpoints. For what it is worth, I may state my
philosophical bias as follows: The most striking fact we have learned
about life as it now exists is the ubiquity of dual structure, the division of every organism into hardware and software components,
into protein and nucleic acid. I consider dual structure to be prima
facie evidence of dual origin. If we admit that the spontaneous emer-
gence of protein structure and nucleic acid structure out of molecular chaos is unlikely, it is easier to imagine two unlikely events occurring separately over a long period than to imagine two unlikely

events occurring simultaneously. Needless to say, vague arguments of this sort, invoking probabilities we are unable to calculate quantitatively, cannot be conclusive. The main reason I am hopeful for progress in the understanding of the origin of life is that the subject is moving away from the realm of philosophical speculation and into the realm of experimental science.

EIGEN AND ORGEL

The third and fourth names on my list of illustrious predecessors are those of Manfred Eigen and Leslie Orgel. Unlike Schrödinger and von Neumann, they are experimenters. They are explorers of experimental approaches to the problem of the origin of life. They are, after all, chemists, and this is a job for chemists. Eigen and his colleagues in Germany did experiments that showed us biological organization originating spontaneously and evolving in a test tube (Fig. 1). More precisely, they demonstrated that a solution of nucleotide monomers will, under suitable conditions, give rise to a nucleic acid polymer molecule that replicates and mutates and competes with its progeny for survival. From a certain point of view, one might claim that these experiments already achieved the spontaneous generation of life from nonlife. They brought us at least to the point where we could ask and answer questions about the ability of nucleic acids to synthesize and organize themselves (Eigen et al., 1981). Unfortunately, the conditions in Eigen's test tubes were not really prebiotic. To make his experiments work, Eigen put into the test tubes a polymerase enzyme, a protein catalyst extracted from a living bacteriophage. The synthesis and replication of the nucleic acid were dependent on the structural guidance provided by the enzyme. We are still far from an experimental demonstration of the appearance of biological order without the help of a biologically derived precursor. Nevertheless, Eigen provided tools with which experimenters may begin to attack the problem of origins.

Leslie Orgel, like Manfred Eigen, is an experimental chemist. He taught me most of what I know about the chemical antecedents of life. He did experiments complementary to those of Eigen. Eigen

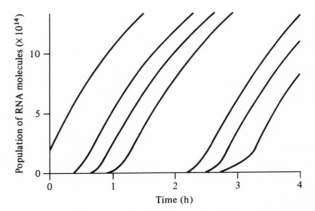

Figure 1 The Biebricher–Eigen–Luce experiment demonstrating evolution of RNA molecules in a test tube containing a solution of nucleotide monomers with added replicase enzyme. The four curves on the left were obtained with 10^{14}, 10^6, 10^3, and 1 molecules of RNA template added to the mixture. The three curves on the right are three separate runs with no template added. (Data from Eigen et al., 1981).

was able to make RNA grow out of nucleotide monomers without having any RNA template for the monomers to copy but with a polymerase enzyme to tell the monomers what to do. Orgel did equally important experiments in the opposite direction. Orgel demonstrated that nucleotide monomers will, under certain conditions, polymerize to form RNA if they are given an RNA template to copy without any polymerase enzyme. Orgel found that zinc ions in the solution are a good catalyst for the RNA synthesis. It may not be entirely coincidental that many modern biological enzymes have zinc ions in their active sites. To summarize, Eigen made RNA using an enzyme but no template, and Orgel made RNA using a template but no enzyme. In living cells, RNA is made using both templates and enzymes. If we suppose that RNA was the original molecule with which life began, then to understand the origin of life we have to make RNA using neither a template nor an enzyme. Neither Eigen nor Orgel came close to achieving this goal.

The belief that life began with RNA, already widely accepted at the time when Eigen and Orgel were doing their experiments, received a strong boost from the discovery of ribozymes by Cech. If, as Cech demonstrated, RNA can perform the function of an enzyme, catalyzing chemical reactions in a primitive cell, then protein enzymes might be unnecessary. Primitive cells might have carried out all the functions of metabolism and replication with RNA alone. The phrase "The RNA World" was introduced (Gilbert, 1986; Joyce, 1989) to describe the state of affairs in early times when RNA-life was evolving without the help of protein enzymes. The experiments of Eigen were extended (Wright and Joyce, 1997) to demonstrate that an RNA ribozyme in the test tube can evolve in such a way as to increase its effectiveness as a catalyst by a factor of ten thousand or more. A very feeble ribozyme evolved into a highly efficient ribozyme in an experiment lasting only five days. In another remarkable experiment (Santoro and Joyce, 1997), molecules of DNA were artificially evolved in test tubes to perform the functions of an enzyme, and the resulting DNA enzyme was even more efficient than the best RNA ribozyme. DNA is a magic molecule with extraordinary properties, and it may have many functions in the cell besides carrying genetic information. However, the experiments of Santoro and Wright and Joyce, like the experiment of Eigen, still required protein enzymes in the test tube. Without polymerase and reverse transcriptase enzymes, the experiments would not work. Ribozymes have not yet been seen to evolve in a test tube containing RNA alone.

I do not consider the existence of ribozymes to be a decisive reason to believe in the existence of an RNA world. Before the discovery of ribozymes, we already knew that RNA performs important hardware functions in addition to its software functions. The ribozyme is only one more item to add to the list of RNA hardware functions. In every one of its hardware functions, as transfer RNA, as ribosomal RNA, or as a ribozyme, RNA is working as part of a machine largely made up of proteins. When I look at the experiments of Eigen and Orgel and Wright and Joyce, I see nothing that resembles an RNA world. I see these experiments fitting more naturally into the framework of a double-origin hypothesis. According to the double-origin hypothesis, RNA was not the original molecule of life. In this hypothesis the original molecules of life were proteins, or polymers similar to proteins, and life of a sort was already established before RNA came into the picture. In this context the Eigen and Orgel and Wright and Joyce experiments are exploring the evolution of RNA under conditions appropriate to the second origin of life. They come close to describing a parasitic development of RNA life within an environment created by a preexisting protein life. Concerning the first origin of life, the origin of protein life and of protein metabolism, they say nothing. The origin of metabolism is the next great virgin territory waiting for experimental chemists to explore.

MARGULIS

The fifth name on my list of illustrious predecessors is that of Lynn Margulis. Although she is still very much alive and considerably younger than I am, she set the style in which I came to think about early evolution. Her style is well displayed in the popular book (Margulis and Sagan, 1995) that portrays the prodigality of life and the mysteries of its evolution in a glowing symbiosis of prose and pictures. She describes the sciences of physiology and ge-netics as two solid foundations of knowledge with a wide river of ignorance running between them. Because we have solid ground on the two sides, we can use our understanding of life's history and evolution to build a bridge over the river. In science a bridge is a theory. When bridges are to be built, theoretical scientists may have a useful role to play.

Lynn Margulis is one of the chief bridgebuilders in modern biol-ogy. She built a bridge between the facts of cellular anatomy and the facts of molecular genetics. Her bridge was the idea that parasitism and symbiosis were the driving forces in the evolution of cellular complexity. She did not invent this idea, but she was its most active promoter and systematizer. The idea was called "symbiogenesis" by its original author, the Russian botanist Konstantin Merezhkovsky (Merezhkovsky, 1909; Khakhina, 1992; Dyson, 1997). It remained popular in Russia but had little support outside Russia until Margulis revived it. She collected the evidence to support her view that the main internal structures of eucaryotic cells did not originate within the cells but are descended from independent living creatures that invaded the cells from outside like carriers of an infectious disease (Margulis, 1970, 1981). The invading creatures and their hosts then gradually evolved into a relationship of mutual dependence. The erstwhile disease organism became by degrees a chronic parasite, a symbiotic partner, and finally an indispensable part of the substance of the host. This Margulis picture of early cellular evolution now has incontrovertible experimental support. The molecular structures of chloroplasts and mitochondria are found to be related more closely to alien bacteria than to the cells in which they have been incorporated for one or two billion years.

In addition, there are general philosophical reasons for believing that the Margulis picture will be valid, even in cases where it cannot be experimentally demonstrated. A living cell, to survive, must be intensely conservative. It must have a finely tuned molecular organization, and it must have efficient mechanisms for destroying promptly any molecules that depart from the overall plan. Any new structure arising within this environment must be an insult to the integrity of the cell. Almost by definition, a new structure will be a disease that the cell will do its best to resist. It is possible to imagine new structures arising internally within the cell and escaping its control like a cancer growing in a higher organism. But it is much easier to imagine new structures coming in from the outside like infectious bacteria already prepared by the rigors of independent living to defend themselves against the cell's efforts to destroy them.

The main reason I find the two-origin hypothesis philosophically congenial is that it fits well into the general picture of evolution portrayed by Margulis. According to Margulis, most of the big steps in cellular evolution were caused by parasites. The double-origin hypothesis implies that nucleic acids were the oldest and most successful cellular parasites. It extends the scope of the Margulis picture of evolution to include not only eucaryotic cells but procaryotic cells as well. It proposes that the original living creatures were cells with a metabolic apparatus directed by enzymes (molecules similar to proteins) but with no genetic apparatus. Such cells would lack the capacity for exact replication but could grow, divide, and reproduce themselves in an approximate statistical fashion. They might have continued to exist for millions of years, gradually diversifying and refining their metabolic pathways. Among other things, they discovered how to synthesize adenosine triphosphate (ATP), the magic molecule that serves as the principal energy-carrying intermediate in all modern cells. Cells carrying ATP were able to function more efficiently and prevailed in the Darwinian struggle for existence. In time it happened that cells were full of ATP and other related molecules such as adenosine monophosphate (AMP).

Now we observe the strange fact that the two molecules, ATP and AMP, which have almost identical chemical structures (Fig. 2), have totally different but equally essential functions in modern cells. ATP is the universal energy carrier. AMP is one of the nucleotides that make up RNA and function as bits of information in the genetic apparatus. To get from ATP to AMP, all you have to do is remove two phosphate moieties. I am proposing that the primitive cells had no genetic apparatus but were saturated with molecules like AMP as a by-product of the energy-carrying function of ATP. This was a dangerously explosive situation, and in one cell that happened to be carrying an unusually rich supply of nucleotides, an accident occurred. The nucleotides began doing the Eigen experiment of RNA synthesis three billion years before it was done by Eigen. Within the cell, with some help from preexisting enzymes, the nucleotides produced an RNA molecule, which then continued to replicate itself. In this way RNA first appeared as a parasitic disease within the cell. The first cells in which the RNA disease occurred probably became sick and died. But then, according to the Margulis scheme, some of the infected cells learned how to survive the infection. The protein-based life learned to tolerate the RNA-based life. The parasite became a symbiont. And then, very slowly over millions of years, the protein-based life learned to make use of the capacity for exact replication that the chemical structure of RNA provided. The primal symbiosis of protein-based life and parasitic RNA grew gradually into a harmonious unity, the modern genetic apparatus.

This view of RNA as the oldest and most incurable of our parasitic diseases is only a poetic fancy, not yet a serious scientific theory.

Adenosine
triphosphate

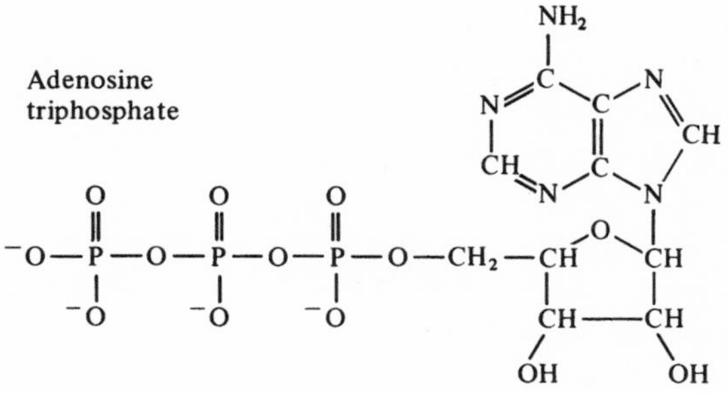

Adenosine
5′-monophosphate

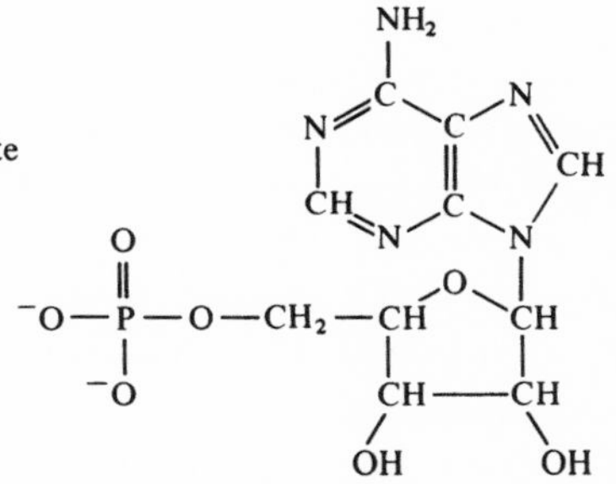

Figure 2 The molecular structures of adenosine triphosphate (ATP) and adenosine 5′-monophosphate (AMP), otherwise known as adenine nucleotide.

Still, it is attractive to me for several reasons. First, it is in accordance with our human experience that hardware should come before software. The modern cell is like a computer-controlled chemical factory in which proteins are hardware and nucleic acids, with the exceptions already mentioned, are software. In the evolution of machines and computers, we always developed the hardware first before we began to think about software. I find it reasonable that natural evolution should have followed the same pattern. A second argument in favor of the parasite theory of RNA comes from the chemistry of amino acids and nucleotides. It is easy to synthesize amino acids, the constituent parts of proteins, out of plausible prebiotic materials. The synthesis of amino acids from a hypothetical reducing atmosphere was demonstrated in a classic experiment by Miller in 1953. Although it is now considered unlikely that the earth ever had a reducing atmosphere, there must always have been local environments in which reducing conditions existed. In particular, the existence of amino acids in some ancient meteorites proves that prebiotic synthesis of amino acids is possible. The nucleotides that make up nucleic acids are much more difficult to synthesize. Nucleotide bases such as adenine and guanine have been synthesized by Oró from ammonia and hydrocyanic acid. But to go from a base to a complete nucleotide is a more delicate matter. Furthermore, once formed, nucleotides are less stable than amino acids. Because of the details of the chemistry, it is much easier to imagine a droplet of water on the prebiotic earth becoming a rich soup of amino acids than to imagine a droplet becoming a rich soup of nucleotides. Charles Darwin imagined life beginning in a "warm little pond" on the surface of the earth. Recently Thomas Gold and others (Gold, 1992, 1998; Chyba and McDonald 1995) have suggested that a hot, deep environment is a more likely birthplace for life. In either case, nucleotides would be difficult to make and easy to destroy. Nucleotides would have had a better chance to accumulate and polymerize if they originated in biological processes inside the protective environment of already-existing cells.

My third reason for preferring the parasite theory of RNA is that it may be experimentally testable. If the theory is true, living cells may have existed for a long time before becoming infected with nucleic acids. There exist microfossils, traces of primitive cells, in rocks that are more than three billion years old. It is possible that some of these microfossils might come from cells older than the origin of RNA. It is possible that the microfossils may still carry evidence of the chemical nature of the ancient cells. For example, if the microfossils were found to preserve in their mineral constituents significant quantities of phosphorus, this would be strong evidence that the ancient cells already possessed something resembling a modern genetic apparatus. So far as I know, no such evidence has been found. I do not know whether the processes of fossilization would be likely to leave chemical traces of nucleic acids intact. So long as this possibility exists, we have the opportunity to test the hypothesis of a late origin of RNA by direct observation.

KIMURA

The last of the illustrious predecessors on my list is the geneticist Motoo Kimura, who died in 1994 on his seventieth birthday. Kimura developed the mathematical basis for a statistical treatment of molecular evolution (Kimura, 1970), and he has been the chief advocate of the neutral theory of evolution (Kimura, 1983). The neutral theory says that, through the history of life from beginning to end, random statistical fluctuations have been more important than Darwinian selection in causing species to evolve. Evolution by random statistical fluctuation is called genetic drift. Kimura maintains that genetic drift drives evolution more powerfully than natural selection. I am indebted to Kimura in two separate ways. First, I use Kimura's mathematics as a tool for calculating the behavior of molecular populations. The mathematics is correct and useful whether you believe in the neutral theory of evolution or not. Second, I find the neutral theory of evolution helpful even though I do not accept it as dogma. In my opinion, Kimura has overstated his case, but still his picture of evolution may sometimes be right. Genetic drift and natural selection are both important, and there are times and places where one or the other may be dominant. In particular, I find it reasonable to suppose that genetic drift was dominant in the very earliest phase of biological evolution before the mechanisms of heredity had become established. Even if the neutral theory is not true in general, it may be a useful approximation to make in building models of prebiotic evolution.

We do not know whether the origin of life was gradual or sudden. It might have been a process of slow growth stretched out over millions of years, or it might have been a single molecular event that happened in a fraction of a second. As a rule, natural selection is more important over long periods, and genetic drift is more important over short periods. If you think of the origin of life as being slow, you must think of it as a Darwinian process driven by natural selection. If you think of it as being quick, then the Kimura picture of evolution by statistical fluctuation without selection is appropriate. In reality the origin of life must have been a complicated process with incidents of rapid change separated by long periods of slow adaptation. A complete description needs to take into account both drift and selection. In my calculations I have made use of the theorist's privilege to simplify and idealize a natural process. I have considered the origin of life to be an isolated event occurring on a rapid timescale. In this hypothetical context, it is consistent to examine the consequences of genetic drift acting alone. Darwinian selection will begin its work after the process of genetic drift has given it something to work on.

If one wishes to examine seriously the double-origin hypothesis, the hypothesis that life began and flourished without the benefit of exact replication, then it is appropriate to assume that genetic drift remained strong and natural selection remained relatively weak during the early exploratory phases of evolution. But this is not to say that Darwinian selection had to wait until life learned to replicate exactly. Darwinian selection is not logically dependent on exact replication. Indeed, Darwin himself knew nothing of exact replication when he invoked the idea of natural selection. Darwinian selection would have operated to guide the evolution of living creatures, even at a time when those creatures may have lacked anything resembling a modern genetic apparatus. All that is necessary for natural selection to operate is that there be some inheritance of chemical constituents from an organism to its progeny. The inheritance need not be exact. It is sufficient if a cell splitting into two daughter cells transmits to each of its daughters with high probability a population of molecules capable of continuing its own pattern of metabolism. Darwinian selection is unavoidable as soon as inheritance begins, no matter how sloppy the mechanism of inheritance may be. When I apply Kimura's mathematics of genetic drift to describe the earliest phase of the first origin of life, this does not mean that I am following Kimura in his belief that genetic drift continued to be dominant later. I consider it unlikely that genetic drift continued to be dominant once life was well established, even if the early forms of life were incapable of exact replication.

There are many other illustrious predecessors besides those whom I have mentioned. I chose to speak of these six – Schrödinger, von Neumann, Eigen, Orgel, Margulis, and Kimura – because each of them is in some sense a philosopher as well as a scientist. Each of them brought to biology not just technical skills and knowledge but a personal philosophical view-point extending beyond biology over the whole of science. From all of them I have borrowed the ideas that fitted together to form my own philosophical viewpoint. The origin of life is one of the few scientific problems broad enough to make use of ideas from almost all scientific disciplines. Schrödinger brought to it ideas from physics, von Neumann ideas from mathematical logic, Eigen and Orgel ideas from chemistry, Margulis ideas from ecology, and Kimura ideas from population biology. What I am trying to do in this book is to explore the connections, to see whether mathematical logic and population biology may have raised new questions that chemistry may be able to answer.

CHAPTER TWO

Experiments and Theories

The study of prebiotic evolution divides itself into three main stages, which one may label geophysical, chemical, and biological. The geophysical stage concerns itself with the early history of the earth and especially with the nature of the earth's primitive crust, ocean, and atmosphere. The chemical stage concerns itself with the synthesis, by natural processes operating within plausible models of the primitive atmosphere and ocean, of the chemical building blocks of life. When we speak of building blocks, we tend to think of the amino acids and nucleotide monomers out of which the proteins and nucleic acids in modern cells are built. The building blocks at the beginning of life may have been very different, but the majority of experiments exploring prebiotic synthesis have been aimed at the synthesis of amino acids and nucleotides. The biological stage concerns itself with the appearance of biological organization, with the building of a coordinated population of large molecules with catalytic functions out of a random assortment of building blocks. If the building blocks were amino acids or nucleotides, the large molecules would have been proteins or nucleic acids. But biological organization probably began with a far more heterogeneous population of molecules than we see in modern cells.

Generally speaking, it can be said that the geophysical and chemical stages of prebiotic evolution are reasonably well understood. At least these two stages are in the hands of competent experts, and I have nothing significant to add to what the experts have reported. Theories of the geophysical stage can be checked by abundant observations in the field of geology. Theories of the chemical stage can be checked by experiments done by chemists in the laboratory. Many details remain to be elucidated, but the geophysical and chemical stages are no longer shrouded in mystery. I have therefore concentrated my attention on the biological stage. The problem of the origin of life is for me the biological stage, the problem of the appearance of biological organization out of molecular chaos. It is in this biological stage that big mysteries still remain. The purpose of my own work has been to try to define precisely what we mean by the appearance of biological organization and thereby to make the biological stage accessible to experimental study.

CHEMISTRY

The chemical stage of prebiotic evolution was explored in the classic experiment of Miller in 1953 and in many later experiments (Miller and Orgel, 1974). Miller took a reducing atmosphere composed of methane, ammonia, molecular hydrogen, and water; passed electric sparks through it; and collected the reaction products. He found a mixture of organic compounds containing a remarkably high fraction of amino acids. In particular he found a 2-percent yield of alanine. Glycine and alanine are the simplest of the twenty amino acid building blocks out of which all living creatures build proteins. Miller also found that when he added hydrogen sulphide to his atmosphere he obtained respectable yields of the essential sulphur-containing amino acids methionine and cysteine. The experiment worked almost equally well with an atmosphere of molecular hydrogen, molecular nitrogen, and carbon monoxide. It failed completely in an oxidizing atmosphere containing either free oxygen or sulphur dioxide. It failed almost completely in a neutral atmosphere composed of molecular nitrogen and carbon dioxide and water, producing extremely small yields of amino acids. Other people have repeated the Miller experiments with many variations by using ultraviolet light or ionizing radiation as the energy source instead of electric sparks. The results are always consistent. The input of energy into a reducing atmosphere causes production of amino acids in substantial quantity. The input of energy into a neutral or oxidizing atmosphere does not.

The prebiotic synthesis of nucleotides is a more difficult problem. Efforts to synthesize nucleotides directly from their elementary components in a Miller-style experiment have not been successful. A nucleotide is a lopsided molecule made up of three parts: an organic base plus a sugar plus a phosphate ion. The most plausible way to synthesize an organic base was demonstrated in an experiment by Oró in 1960. Oró prepared a concentrated solution of ammonium cyanide in water and simply let it stand. He found that the ammonium cyanide was converted into the organic base adenine with 0.5-percent yield. Adenine is one of the four bases (adenine, thymine, guanine, and cytosine) that are the active ingredients of DNA. Oró was also able to synthesize guanine in a similar way. But the starting material, ammonium cyanide, is unlikely to have been abundant in the surface waters of the earth unless the atmosphere was reducing. The Oró experiment, like the Miller experiment, needs a reducing atmosphere to work well. It is not easy to imagine an ammonium cyanide solution under natural conditions becoming sufficiently concentrated to make the Oró synthesis occur with high yield. Leslie Orgel has suggested one possible way in which this might have happened. If a pond containing a dilute solution of ammonium cyanide freezes, the ice on top will be almost pure water, and the solution in the unfrozen liquid below will become more concentrated as the freezing proceeds. If the temperature falls slowly and the freezing continues smoothly, the final result will be a small volume of concentrated eutectic solution of ammonium cyanide at the bottom of the pond. The temperature at which the eutectic solution finally freezes is −22°C. Conceivably, the concentrated solution at a temperature around −20°C might remain undisturbed for a long enough time to produce adenine by the Oró reaction. As Leslie Orgel has remarked, what we need to give us a natural synthesis of nucleotide bases is not a warm soup but a very cold soup.

The Oró synthesis of nucleotide bases requires much more special conditions than the Miller synthesis of amino acids. The sugar component of nucleotides is also difficult to produce. The sugar can be synthesized with reasonable efficiency from a concentrated solution of formaldehyde. Formaldehyde is a molecule seen occurring naturally in molecular clouds in the sky. But the sugar synthesis, like the Oró synthesis, requires high concentration. And formaldehyde, like ammonium cyanide, prefers a reducing atmosphere. The bases and the sugar are both unlikely to arise spontaneously under neutral conditions. The third component of nucleotides is the phosphate ion. This is the only component that occurs naturally as a mineral constituent in rocks and seawater and does not need to be synthesized.

We have thus found possible, although unlikely, ways for each of the three parts of a nucleotide to occur in a prebiotic environment. Even more severe difficulties arise when we try to find a natural way to stick the three parts together in the right geometrical arrangement. If the linkages are made at random, only about one in a hundred molecules will be stereochemically correct. Only the correctly linked molecules can polymerize to make nucleic acids. It is difficult to imagine a prebiotic process that could separate a correctly formed nucleotide from its ninety-nine misshapen brothers. And finally, the correctly formed nucleotides are unstable in solution and tend to hydrolyze back into their components. We cannot assume that nucleotides continued to accumulate in primitive ponds for thousands of years. The rate of synthesis of nucleotides must be high to keep pace with the rate of hydrolysis. The nucleotides in our bodies are stable only because they are packaged in double helices that protect them from hydrolysis. The nucleotides on the primitive earth would have been rare birds, difficult to synthesize and easy to dissociate. Nobody has yet discovered a way to make them out of their components rapidly enough so that they would have a reasonable chance of finding each other and combining into stable helices before they hydrolyzed.

The results of thirty years of intensive chemical experimentation have shown that the prebiotic synthesis of amino acids is easy to simulate in a reducing environment, but the prebiotic synthesis of nucleotides is difficult in all environments. We cannot say that the prebiotic synthesis of nucleotides is impossible. We know only that, if it happened, it happened by some process that none of our chemists has been clever enough to reproduce. This conclusion may be considered to favor the double-origin hypothesis and argue against a single-origin hypothesis for the origin of life. A single-origin hypothesis requires amino acids and nucleotides to be synthesized by natural processes before life began. The double-origin hypothesis requires only amino acids to be synthesized prebiotically, the nucleotides being formed later as a by-product of protein metabolism. The evidence from chemical simulations does not disprove the single-origin hypothesis but makes a strong presumptive case against it.

GENETICS AND PALEONTOLOGY

I have summarized some evidence about the origin of life provided by chemistry. I now discuss evidence from genetics and paleontology. The main fact that we have learned from genetics is that the genetic apparatus is universal. By the genetic apparatus I mean the organization of ribosomes and transfer RNA molecules that enables a cell to translate a nucleic acid gene into a protein according to the triplet code. In modern cells the gene is transcribed into a molecule of messenger RNA before being translated. The forms of messenger RNA are highly variable from species to species. The triplet code is embodied in the transfer RNA molecules and is the same in all cells apart from some minor differences. This universality of the genetic apparatus is strong evidence that all existing cells are descended from a common ancestor. There must have been a unique latest common ancestor, a single cell whose progeny diversified into the myriad branches of the evolutionary tree. We know from the genetic evidence that the latest common ancestor already possessed an essentially complete genetic apparatus with the same triplet code used in modern cells.

George Fox and Carl Woese were the chief explorers of the earliest branchings of the evolutionary tree. They delineated the early branches by quantitatively measuring the degree of relatedness of the nucleotide sequences of ribosomal RNA molecules in widely diverse cells. The ribosomal RNA molecules, being crucial to the functioning of the genetic apparatus, are intensely conservative and change their sequences extremely slowly. Nevertheless they do change, and the divergence between sequences in two cells measures in a rough fashion the time that has elapsed since those two cells shared a common ancestor. Tracing the relationships between cells in this way, one finds that the evolutionary tree has three main branches representing a divergence of cell types far more ancient than the later division of creatures into animals and plants. Moreover, the genetic apparatus carried by organelles such as chloroplasts and mitochondria within eucaryotic cells does not belong to the same main branch of the tree as the genetic apparatus in the nuclei of the eucaryotic cells. The difference in genetic apparatus between organelles and nucleus is the strongest evidence confirming Lynn Margulis's theory that the organelles of the modern eucaryotic cell were originally independent free-living cells and only later became parasites of the eucaryotic host. According to this theory, the evolutionary success of the eucaryotic cell was due to its policy of free immigration. Like the United States of America in the nineteenth century, the eucaryotic cell gave shelter to the poor and homeless and exploited their talents for its own purposes. Needless to say, both in the United States and in the eucaryotic cell, once the old immigrants are comfortably settled and their place in society is established, they do their best to shut the door to any prospective new immigrants.

There are some differences, not only in the sequences of ribosomal RNA molecules but in the genetic code itself, between mitochondria and independently living cells. The mitochondria of various species have minor variations of the code that are not seen elsewhere. This fact is additional evidence for the parasitic origin of mitochondria. Nevertheless, that differences exist is less important than the fact that differences are very slight. Even in the most striking cases, the mitochondrial code is close to the standard code, and the two codes cannot be genetically unrelated. The exceptions to the universality of the code do not weaken the case for believing that the mitochondrion and its host, for however long a time they may have been separated, were originally descended from a common ancestor.

This argument for the existence of a common ancestor applies only to cells and organelles that possess a genetic translation apparatus. It does not apply to creatures such as viruses that reproduce only within cells and borrow the genetic apparatus of the cells they invade. The genetic structures of viruses do not give us direct evidence of their antiquity. Viruses may be very ancient, or they may have originated comparatively recently as plasmids, that is to say as pieces of nucleic acid detached from the chromosomes of normal cells and pursuing a more or less independent existence within the cells. A virus may be nothing more than a plasmid that has learned to survive outside its host cell by covering itself with a protein coat. Alternatively, a virus may be a highly degenerate descendant of a normal cell that has adopted a parasitic mode of life and lost all of its metabolic functions. The origin of viruses is still an open question. All that we can say for sure is that, because viruses as they now exist are totally parasitic, there must have been cells before there were viruses. There is no way in which we can imagine a virus coming first and later growing into a cell.

We know even less concerning the possible origin of a mysterious group of organisms that have been given the name of "prions." These are the organisms that are responsible for some slow degenerative diseases of the central nervous system: scrapie in sheep, kuru and Creutzfeldt–Jakob disease in humans, and most notoriously, bovine spongiform encephalopathy in cows. When these diseases were first identified and studied, the agents causing them were assumed to be viruses. Carleton Gajdusek told in his Nobel Prize lecture (Gajdusek, 1977) the dramatic story of kuru, the disease that almost exterminated the Fore tribe in New Guinea. The Fore people made a habit of eating the brains of members of the tribe who died. Gajdusek stopped the epidemic by persuading them that human brains were bad for their health. But all attempts to find the kuru virus failed. Stanley Prusiner (Prusiner, 1982, 1991) worked for many years on the chemical analysis of the scrapie agent and concluded that it could not be a virus. He tentatively identified the agent as a modified form of a known protein. He gave it the name "prion," meaning proteinaceous infective particle. But all that is known for certain is that the agent multiplies and causes disease

within the brains of animals and that it is unique among living creatures in giving no positive response to any of the standard chemical
tests for the presence of nucleic acid.   It is conjectured, but not
proved, that the prion is pure protein without any nucleic acid.
How it might succeed in reproducing without nucleic acid is still a
mystery. Presumably it has found a way to invade nerve cells and
induce the protein in the cells to change into copies of itself. For
this to be possible, it must be made of the same kind of protein that
exists naturally in nerve cells. The structure and life cycle of prions
are now being actively investigated in many places, and with luck
within a few years we shall understand prions as well as we understand viruses. The understanding of their structure may or may not
lead to an understanding of their origin.

I am putting forward in this book such evidence as I can collect
to support the hypothesis that life had a double origin. The double-
origin hypothesis implies that the first living creatures were able to
metabolize but not to replicate and that they were built of molecules
resembling proteins rather than nucleic acids. You might at this
point expect me to claim that the existence of the prion is evidence
confirming this hypothesis. I make no such claim. The prion would
be confirming evidence only if it could be proved to be a primeval
relic directly descended from the earliest creatures that lived before
the development of the nucleic acid genetic apparatus. There are
several strong reasons that convince me, much as I would like to
discover such a relic, that the prion cannot be primeval. The prion
is like a virus in having a narrowly specialized and wholly parasitic
life cycle. Even more than a virus, it must be closely linked in chemical structure to the cells it invades. It is difficult to imagine that a
primeval cell, after lurking in odd corners of the earth for billions of
years, would miraculously find itself preadapted to the chemistry of
so sophisticated an organ as the brain of a sheep. It is far more likely
that the prion originated in modern times as a displaced fragment of
a brain cell. Another argument against the antiquity of prions is the
improbability of survival of creatures without a genetic apparatus
in competition with creatures of modern design. If cells without a
genetic apparatus did indeed exist, they cannot have had metabolic
agility, mobility, and responsiveness in any way comparable with

modern cells. They must have been grossly inefficient, slow, and blind by modern standards. As soon as the genetic apparatus was perfected, cells possessing it had an overwhelming advantage over earlier forms of life. No matter how long the evolution of nonreplicating forms of life may have lasted, and no matter how great a variety of such forms may have existed, we should not expect any living relics of that epoch to have survived. The prion is an exciting and important discovery, but it is unlikely to throw any direct light upon the question of the origin of life.

The evidence of genetic relatedness collected by Carl Woese and others proves that all existing cells have a common ancestor but does not provide an absolute date for the epoch of the latest common ancestor. Genetic evidence gives us good relative dating of the different branches of the evolutionary tree but no absolute dating. For absolute dates we must turn to the evidence of paleontology. The pioneer in the discovery of fossil evidence for the absolute dating of early life was E. S. Barghoorn. The rock in which microfossils are best preserved is chert, the geologists' name for the fine-grained silica rock that ordinary people call flint. Chert is formed by the slow precipitation of dissolved silica from water, a process that puts minimal stress on any small creatures that may become embedded within it. Once formed, the chert is hard and chemically inert so that fossils inside it are well protected. The microfossils Barghoorn and others have collected are little black blobs in which internal structure is barely discernible. Not all of them are definitely known to be organic in origin. I myself cannot pretend to decide whether a microscopic blob is a fossil cell or an ordinary grain of dust. I accept the verdict of the experts who say that most of the blobs are in fact fossils.

The results of a great number of observations of microfossils can be briefly summarized as follows. I use the word eon to mean a billion years. In rocks that are reliably dated with age about 3 eons, mainly in South Africa, we find microfossils that resemble modern bacteria in shape and size. In rocks dated with age about 2 eons, mainly in Canada, we find fossils that resemble modern procaryotic algae, including chains of cells and other multicellular structures. In rocks dated with age about 1 eon, mainly in Australia, we find fossils that resemble modern eucaryotic cells with some traces of internal structure. The fossils are too small to be analyzed chemically with any accuracy, but traces of long-chain hydrocarbons have been found in the 3-eon group, whereas the 1-eon group contains porphyrin residues that are presumably derived from chlorophyll.

The geological dating of the various fossil groups is remarkably accurate and reliable. Unfortunately, we do not know with equal accuracy what it is that is being dated. We do not know how to identify the various fossils with particular branches of the evolutionary tree. Except for the general similarity of size and shape, there is no evidence that the cells of the 3-eon group were cousins of modern bacteria. There is no evidence that they possessed a modern genetic apparatus with ribosomes and transfer RNA. There is no evidence of the presence of nucleic acids in any of the ancient microfossils. So far as the evidence goes, the cells of the 3-eon group may have been either cells of modern type with a complete genetic apparatus or cells of a rudimentary kind lacking nucleic acids altogether, or anything in between. Only the cells of the 1-eon group were definitely modern with eucaryotic characteristics. So far as the evidence goes, the latest common ancestor of all living cells may have lived before the 3-eon group of fossils, or between the 3-eon group and the 2-eon group, or possibly even later than the 2-eon group. The dating of the latest common ancestor requires a reliable linkage of the branchpoints of the evolutionary tree with the various groups of fossils. The most urgent problem for evolutionary geneticists and paleontologists is to establish the calibration of relative ages determined by genetic linkages in terms of absolute ages determined by geology. Until this problem is solved, neither the genetic evidence nor the paleontological evidence will be sufficient to determine the date of our latest common ancestor.

The interval of time between the beginning of life and the latest common ancestor may have been very long. Some new geological evidence has upset preconceived notions and raised new questions about the date of life's origin. The new evidence comes from the most ancient of all known terrestrial rocks in Greenland. The Greenland rocks are reliably dated and are at least 3.8 eons old. They contain tiny carbonaceous inclusions that must be at least as old as the rocks. Careful analysis with microprobes of the abundances of carbon isotopes in the inclusions shows that the carbon thirteen isotope is depleted to a degree characteristic of biologically processed carbon (Mojzsis et al., 1996). This suggests that life existed on earth very soon after the time of heavy bombardment when the lunar highlands became densely cratered. The ages of rocks brought back by the Apollo astronauts from the moon have been reliably determined. They show that massive impacts were occurring on the moon until 3.8 eons ago. Equally massive impacts must have been occurring on the earth at the same time. Nobody expected life on earth to be established so early, but the evidence in the Greenland rocks has to be taken seriously. The orthodox view until recently was that life originated on earth some time between 3.5 and 3.8 eons ago. The new evidence from Greenland suggests that life may be more ancient than we supposed. It appears to have been spread widely over the earth even before the era of heavy bombardment ended.

The Deep Hot Biosphere

At the time when Miller did his experiments, and for many years afterwards, the prevailing opinion among the experts was that the atmosphere of the primitive earth was reducing. This belief was based on astronomical evidence. Radio astronomers using millimeter-wave telescopes observed that our galaxy is thickly populated with molecular clouds containing large quantities of molecular hydrogen, water, ammonia, carbon monoxide, methyl alcohol, hydrocyanic acid, and other molecules, all of them reducing rather than oxidizing. Observation also showed that these molecular clouds are the places where stars are at present being formed by gravitational condensation of the molecule-bearing gas. Presumably, the earth and sun were formed by condensation of a similar molecular cloud 4.5 eons ago. It seemed reasonable to suppose that the primitive earth contained enormous quantities of the same reducing molecules that we see in the molecular clouds today. The original atmosphere of the earth might have been stripped off many times by intense radiation from the sun or by the violent infall of planetesimals as the earth accumulated. Nevertheless, the atmosphere that remained after the solar system emerged into quiescence from the initial violence was believed to have been mainly composed of the molecules that exist so abundantly in the sky. The experts believed that at the epoch of life's origin the earth's atmosphere was reducing and contained the hydrogen-rich species ammonia, methane, and molecular hydrogen that Miller used in his experiment. Miller's experiment was supposed to be a true simulation of prebiotic chemistry on the primitive earth. But now nobody believes this any more.

Two lines of evidence have made it clear that the reducing atmosphere, if it ever existed, had disappeared by the time the heavy meteoritic bombardment of the earth ceased about 3.8 eons ago. First, there is the direct evidence from geology. Sedimentary rocks laid down on the ancient earth, including carbonates and various oxidized forms of iron, have been reliably dated with ages going all the way back to 3.8 eons. These rocks could not have formed under reducing conditions. Their composition proves that the atmosphere was neutral from 3.8 until about 2 eons ago, when molecular oxygen first appeared. The atmosphere became oxidizing after life was well established and photosynthetic organisms began to produce free oxygen in large quantities.

The second line of evidence for an early neutral atmosphere comes from the rarity of the inert gases remaining in the earth's atmosphere today. Neon is the seventh most abundant element in the universe and is abundant in the molecular clouds out of which the earth condensed. If any primitive reducing atmosphere had survived after the heavy bombardment stopped, it should have contained a large fraction of neon. Neon should have been about as abundant as nitrogen. When the hypothetical reducing atmosphere later became neutral or oxidizing, the neon would have remained. But today the ratio of neon to nitrogen in the atmosphere is one to sixty thousand. The nitrogen-rich and neon-poor atmosphere must have been produced, probably by volcanoes discharging gases from the earth's interior, long after any primitive reducing atmosphere was swept away. The reducing atmosphere may well have been swept away before the earth was born. When the earth first condensed out of a molecular cloud, the cloud was already differentiated into gas and dust. The dust, containing the nonvolatile molecules, mostly silicates and metals, with some fraction of graphite and ice, condensed to form the earth. The gas, containing the neon, molecular hydrogen, methane, and ammonia, all the reducing molecules that Miller needed for his experiment to work, did not condense in the neighborhood of the earth but moved out away from the sun to form Jupiter and the planets beyond. The rarity of neon proves that the earth's atmosphere since life began had nothing to do with the gases that we see in molecular clouds in the sky. The atmosphere when life began was neutral. Miller's experiment only shows what might have happened if circumstances had been different.

Since Miller's beguiling picture of a pond full of dissolved amino acids under a reducing atmosphere has been discredited, a new beguiling picture has come to take its place. The new picture has life originating in a hot, deep, dark little hole on the ocean floor. Four experimental discoveries came in rapid succession to make the new picture seem plausible. The first discovery is the presence of abundant life today around vents on the midocean ridges several kilometers below the surface, where hot water emerging from deep below is discharged into the ocean. The water entering the ocean is saturated with hydrogen sulphide and metallic sulphides, and provides a reducing environment independent of the atmosphere above. The second discovery is that bacterial life exists today in strata of rock deep underground, in places where contact with surface life is unlikely. The third finding is the strikingly lifelike phenomena observed in the laboratory, when hot water saturated with soluble iron sulphides is discharged into a cold water environment (Russell et al., 1994). The sulphides precipitate as membranes and form gelatinous bubbles. The bubbles look like possible precursors of living cells. The membrane surfaces adsorb organic molecules from solution, and the metal sulphide complexes catalyze a variety of chemical reactions on the surfaces. The fourth piece of evidence is the discovery that a majority of the most ancient lineages of bacteria are thermophilic, that is to say, specialized to live and grow in hot environments (Nisbet, 1995). The ancient lineages were identified by compiling sequences of bases in riboso-mal RNA of many species and using the observed similarities and differences of the sequences to construct a phylogenetic tree. The phylogenetic tree has a root that represents the hypothetical RNA sequence of the ribosomes in the latest common ancestor of all life. The most ancient lineages are those that branch off closest to the root of the tree. They are found today predominantly in hot springs, often in places where the water temperature is close to boiling.

These four lines of evidence, from ocean ridges, from deep oil-well drilling, from laboratory experiments, and from genetic anal-ysis combine to make the picture of life originating in a hot, deep environment credible. Because we know almost nothing about the origin of life, we have no basis for declaring any possible habitat for life to be likely or unlikely until we have explored it. The picture of life beginning in a deep hot crevice in the earth is purely spec-ulative and in no sense proved. It has an important corollary. If it is true, it implies that the origin of life was largely independent of conditions on the surface of the planet. And this in turn implies that life might have originated as easily on Mars as on the earth (Gold, 1992, 1998). Thomas Gold postulates a deep, hot biosphere still ex-isting in the crust of the earth. He presents evidence that the deep biosphere may contain as much biomass as the surface biosphere with which we are familiar. In many places where samples of rock from deep drilling have been examined, both on land and under the oceans, living bacteria are found in the rock. In many cases the bacteria do not belong to any known species. It is unlikely that their presence in the rock could have resulted from contamination during the process of drilling. Gold remarks, "If in fact such life orig-inated at depth in the earth, there are at least ten other planetary bodies in our solar system that would have had a similar chance for originating microbial life." I do not know which ten objects he has in mind. Certainly Mars, Europa, Titan, and Triton would be on the list. Mars and Europa are prime targets for space missions searching for traces of extraterrestrial life.

THEORIES

We have experimental and observational evidence concerning things that happened before and after the origin of life. Before the origin of life there were geophysical and chemical processes that left behind traces observable in the earth's rocks and in the sky. After the origin of life there were evolutionary processes, which can be observed in fossils and in the taxonomy of nucleic acid molecules in living species. Concerning the origin of life itself, the watershed between chemistry and biology, the transition between lifeless chemical activity and organized biological metabolism, there is no direct evidence at all. The crucial transition from disorder to order left behind no observable traces. When we try to understand the nature of this transition we are forced to go beyond experimental evidence and take refuge in theory. I will now summarize the efforts that have been made by various people over the last eighty years to understand the transition theoretically.

Oparin

There are three main groups of theories about the origin of life. I call them after the names of their most famous advocates: Oparin, Eigen, and Cairns-Smith. I have not done the historical research that would be needed to find out who thought of them first. The Oparin theory was described in Oparin's book *The Origin of Life* in 1924 long before anything was known about the structure and chemical nature of genes (Oparin, 1957). Later in his life, Oparin became a powerful figure in the Soviet Academy of Sciences during the years when Lysenko was suppressing research in genetics in the Soviet Union. Lysenko was a plant breeder who considered genetics to be an obstacle to the fulfilment of Soviet agricultural plans. Stalin gave him dictatorial power, which he used to silence his opponents and to put many of them in prison. Oparin was a friend of Lysenko and did nothing to help the geneticists who were being persecuted (Jukes, 1997). But his moral delinquencies do not necessarily imply that his theory is wrong. Oparin supposed that the order of events in the origin of life was cells first, enzymes second, and genes third.

He observed that when a suitably oily liquid is mixed with water it sometimes happens that the two liquids form a stable mixture called a coacervate with the oily liquid dispersed into small droplets that remain suspended in the water. Coacervate droplets are easily formed by nonbiological processes, and they have a certain superficial resemblance to living cells. Oparin proposed that life began by the successive accumulation of more and more complicated molecular populations within the droplets of a coacervate. The physical framework of the cell came first provided by the naturally occurring droplet. The enzymes came second, organizing the random population of molecules within the droplet into self-sustaining metabolic cycles. The genes came third because Oparin had only a vague idea of their function, and they appeared to him to belong to a higher level of biological organization than enzymes.

One of the proponents of the Oparin theory today is Doron Lancet, who runs computer simulations of the origin of life at the Weizmann Institute in Israel (Segré and Lancet, 1999). He explores a world I like to call the "garbage-bag world." It is the antithesis of the RNA world that the majority of molecular biologists believe in. The RNA world is a neat and beautiful scene with busy little ribozymes cooperating to organize the beginnings of life. The garbage-bag world is not so elegant and not so widely accepted. It is a generalized version of the world imagined by Oparin. Life began with little bags, the precursors of cells, enclosing small volumes of dirty water containing miscellaneous garbage. A random collection of molecules in a bag may occasionally contain catalysts that cause synthesis of other molecules that act as catalysts to synthesize other molecules, and so on. Very rarely a collection of molecules may arise that contains enough catalysts to reproduce the whole population as time goes on. The reproduction does not need to be precise. It is enough if the catalysts are maintained in a rough statistical fashion. The population of molecules in the bag is reproducing itself without any exact replication. While this is happening, the bag may be growing by accretion of fresh garbage from the outside, and the bag may occasionally be broken into two bags when it is thrown around by turbulent motions. The critical question is then, what is the probability that a daughter bag produced from the splitting of a

bag with a self-reproducing population of molecules will itself contain a self-reproducing population? When this probability is greater
than one half, a parent produces on the average more than one
functional daughter, a divergent chain reaction can occur, the bags
containing self-reproducing populations will multiply, and life of a
sort has begun.

The life that begins in this way is the garbage-bag world. It is a
world of little protocells that only metabolize and reproduce themselves statistically. The molecules that they contain do not replicate
themselves exactly. Statistical reproduction is a good enough basis for natural selection. As soon as the garbage-bag world begins
with crudely reproducing protocells, natural selection will operate
to improve the quality of the catalysts and the accuracy of the reproduction. It would not be surprising if a million years of selection
would produce protocells with many of the chemical refinements
that we see in modern cells.

Eigen

The Oparin picture was generally accepted by biologists for half
a century. It was popular not because there was any evidence to
support it but rather because it seemed to be the only alternative
to biblical creationism. Then, during the last forty years, Manfred
Eigen provided an alternative by turning the Oparin theory upside
down (Eigen et al., 1981). The Eigen theory reverses the order of
events. It has genes first, enzymes second, and cells third. In the beginning, in the gospel according to Eigen, was the RNA world. This
is now the most fashionable and generally accepted theory. It has
become popular for three reasons. First, the experiments of Eigen
and Orgel use RNA as the working material and made it plausible
that the replication of RNA was the fundamental process around
which the rest of biology developed. Second, the experiments of
Cech showed that RNA could act as an efficient catalyst in the absence of protein enzymes. Third, the discovery of the double helix
showed that genes are structurally simpler than enzymes. Once the
mystery of the genetic code was understood, it became natural to
think of the nucleic acids as primary and of the proteins as secondary

structures. Eigen's theory has self-replicating RNA at the beginning, enzymes appearing soon afterwards to build with the RNA a primitive form of the modern genetic transcription apparatus, and cells appearing later to give the apparatus physical cohesion.

The Eigen theory is based on two concepts that he calls quasi species and hypercycles. The quasi species comes first and is concerned only with replication. The hypercycle comes a short time later and introduces a metabolic system coupled to the replicative system. Eigen supposes that life began with an ample supply of nucleotide monomers capable of polymerizing into RNA as they did in his laboratory experiments. A quasi species is a population of genetically related but not identical RNA molecules. The molecules in a quasi species form templates for the formation of new RNA molecules that also belong to the quasi species. Because the replication process is not perfect, the molecules will mutate and diversify from generation to generation. Nevertheless, Eigen assumes that the variations in molecular structure within a quasi species remain bounded and settle down to a steady state. The population of molecules in a quasi species shares a common morphology, like the population of individual organisms in a biological species. For this common morphology to persist in a steady state, Eigen assumes that there is a Darwinian process of selection favoring the replication of molecules that come close to the quasi-species norm. Eigen describes this situation quantitatively with a simple set of equations representing the balance between Darwinian selection and random errors of replication.

The hypercycle is a higher level of organization that comes into existence when several quasi species of RNA are established and have begun to form chemical associations with friendly populations of protein enzymes. The enzymes associated with one quasi species are supposed to assist the replication of a second quasi species, and vice versa. The linked populations then become locked into a stable equilibrium.

Even at the quasi-species level of the Eigen theory there is a serious difficulty. The central problem for any theory of the origin of replication is that a replicative apparatus has to function almost perfectly if it is to function at all. If it does not function perfectly, it will give rise to errors in replicating itself, and the errors will accumulate from generation to generation. The accumulation of errors will result in a progressive deterioration of the system until it is totally disorganized. This deterioration of the replication apparatus is called the "error catastrophe."

Eigen has given us a simple mathematical statement of the error catastrophe as follows. Suppose that a self-replicating system is specified by N bits of information and that each time a single bit is copied from parent to daughter the probability of error is ϵ. Suppose that natural selection operates to penalize errors by a selection factor S. That is to say, a system with no errors has a selective advantage S over a system with one error, and so on. Then Eigen finds the criterion for survival to be

$$N\epsilon < \log S. \tag{2.1}$$

If the condition (2.1) is satisfied, the selective advantage of the error-free system is great enough to maintain a population with few errors. If the condition (2.1) is not satisfied, the error catastrophe occurs and the replication cannot be sustained. The meaning of this condition (2.1) is easy to interpret in terms of information theory. The left side ($N\epsilon$) of the inequality is the number of bits of information lost by copying errors in each generation. The right side ($\log S$) is the number of bits of information supplied by the selective action of the environment. If the information supplied is less than the information lost in each generation, a progressive degeneration is inevitable.

The condition (2.1) is very stringent. Because the selective advantage of an error-free system cannot be astronomically large, the logarithm cannot be much greater than unity. To satisfy this condition (2.1) we must have an error rate of the order of N^{-1} at most. The condition is barely satisfied in modern higher organisms that have N of the order of 10^8 and ϵ of the order of 10^{-8}. To achieve an error rate as low as 10^{-8}, modern organisms have evolved an extremely elaborate system of double checking and error correction within the replication system. Before any of this delicate apparatus existed, the error rates must have been much higher. The condition

(2.1) thus imposes severe requirements on any theory of the origin of life, which, like Eigen's theory, makes the replication of RNA a central element of life from the beginning.

All the experiments that have been done with RNA replication under abiotic conditions give error rates of the order of 10^{-2} at best. If we try to satisfy the condition (2.1) without the help of preexisting organisms, we are limited to a replication system that can describe itself with less than one hundred bits of information. One hundred bits of information are far too few to describe any interesting catalytic chemistry. This does not mean that Eigen's theory is untenable. It means that Eigen's theory requires an information-processing system that is at the same time extraordinarily simple and extraordinarily free from error. We do not know how to achieve such low error rates in the initial phases of life's evolution. If an RNA world ever existed, it must have lived all the time close to the edge of an error catastrophe.

A penetrating study of the Eigen theory was carried out by Ursula Niesert and her colleagues at Freiburg in Germany (Niesert et al., 1981). The title of their paper is "Origin of Life between Scylla and Charybdis." They ran a large number of computer simulations of quasi species and hypercycles behaving according to Eigen's rules. Niesert discovered that the error catastrophe is not the only catastrophe to which Eigen's molecular populations may succumb. She found three other catastrophes, each occurring with high frequency in her computer runs, and each having an obvious biological interpretation. The three additional catastrophes were called selfish RNA, short circuit, and population collapse. The selfish RNA catastrophe occurs when a single RNA molecule, as a result of a sequence of mutations, learns to replicate itself faster than its competitors but forgets its function as catalyst. The selfish RNA then becomes a parasite and quickly chokes the rest of the population to death. The short-circuit catastrophe occurs when an RNA molecule, which is supposed to be a link in the chain of hypercycle reactions, changes its sequence in such a way as to catalyze a later reaction in the chain. The chain is then short-circuited, and the hypercycle contracts to a simpler hypercycle or to a single cycle. The population collapse catastrophe occurs when, as a result of a statistical fluctuation, the population of molecules in one of the essential components of a hypercycle falls to zero. The entire hypercycle then rapidly collapses.

Niesert found in her computer simulations that the probability of a selfish RNA catastrophe or a short-circuit catastrophe increases with the size of the molecular population. Because a single aberrant molecule causes the catastrophe, the increase of probability with population size is to be expected. In order to escape the selfish-RNA and short-circuit catastrophes for a reasonable length of time, the population of a hypercycle model must be kept small. On the other hand, the probability of the population collapse catastrophe is large for small populations and decreases as the population size increases. Consequently, the hypercycle model must sail carefully between the Scylla of selfish RNA and short circuit and the Charybdis of population collapse. There is only a narrow range of population size for which all three calamities are unlikely, and even at the optimum population size the lifetime of the hypercycle is finite.

Niesert's results are important because they reveal perils likely to exist in any model of the early stages of the development of life. Early life, whether or not it is correctly described by Eigen's hypercycle model, had to sail between the perils of Scylla and Charybdis. Niesert's analysis is not merely a criticism of the Eigen theory. It is a criticism of any theory of the origin of life that assumes a co-operative organization of a large population of molecules without providing explicit safeguards against short-circuiting of metabolic pathways. The Oparin theory has not yet been tested by computer simulations in the style of Niesert. Until it has been so tested, we have no right to assume that it can deal with Niesert's three catastrophes any better than the Eigen theory can.

I chose to study the Oparin theory because it offers a possible way of escape from the error catastrophe. In the Oparin theory, the first living cells had no system of precise replication and could therefore tolerate high error rates. The main advantage of the Oparin theory is that it allows early evolution to proceed in spite of high error rates. It has the first living creatures consisting of populations of molecules with a loose organization and no genetic fine-tuning. There is a high tolerance for errors because the metabolism of the population depends only on the catalytic activity of a majority of the molecules. The system can still function with a substantial minority of ineffective or uncooperative molecules. There is no requirement for unanimity. Because the statistical fluctuations in the molecular populations will be large, there is a maximum opportunity for genetic drift to act as the driving force of evolution.

Cairns-Smith

The third theory of the origin of life, the theory of Cairns-Smith, is based upon the idea that naturally occurring microscopic crystals of the minerals contained in common clay might have served as the original genetic material before nucleic acids were invented (Cairns-Smith, 1982). The microcrystals of clay consist of a regular silicate lattice with a regular array of ionic sites but with an irregular distribution of metals such as magnesium and aluminum occupying the ionic sites. The metal ions can be considered carriers of information like the nucleotide bases in a molecule of RNA. A microcrystal of clay is usually a flat plate with two plane surfaces exposed to the surrounding medium. Suppose that a microcrystal is contained in a droplet of water with a variety of organic molecules dissolved in the water. The metal ions embedded in the plane surfaces form irregular patterns of electrostatic potential that can adsorb particular molecules to the surfaces and catalyze chemical reactions on the surfaces in ways dependent on the precise arrangement of the ions. In this fashion the information contained in the patterns of ions might be transferred to chemical species dissolved in the water. The crystal might thus perform the same function as RNA in guiding the metabolism of amino acids and proteins. Moreover, it is conceivable that the clay microcrystal can also replicate the information contained in its ions. When the crystal grows by accreting silicate and metal ions from the surrounding water, the newly accreted layer will tend to carry the same pattern of ionic charges as the layer below it. If the crystal is later cut along the plane separating the old from the new material, we will have a new exposed surface replicating the original pattern. The clay crystal is thus capable in principle of performing both of the essential functions of genetic material – replicating the information that it carries and transferring the information to other molecules. It can do these things in principle. That is to say, it can do them with some undetermined efficiency, which may be very low. There is no experimental evidence to support the statement that clay can act either as a catalyst or as a replicator with enough specificity to serve as a basis for life. Cains-Smith asserts that the chemical specificity of clay is adequate for these purposes. The experiments to prove him right or wrong have not been done.

The Cairns-Smith theory of the origin of life has clay first, enzymes second, cells third, and genes fourth. The beginning of life was a natural clay crystal directing the synthesis of enzyme molecules adsorbed to its surface. Later, the clay and the enzymes learned to make cell membranes and became encapsulated in cells. The cells contained clay crystals performing in a crude fashion the functions performed in a modern cell by nucleic acids. This primeval clay-based life may have existed and evolved for many millions of years. Then one day a cell made the discovery that RNA is a better genetic material than clay. As soon as RNA was invented, the cells using RNA had an enormous advantage in metabolic precision over the cells using clay. The clay-based life was eaten or squeezed out of existence, and only the RNA-based life survived. Cairns-Smith has published a delightful account of his ideas (Cairns-Smith, 1985) in nontechnical language. He uses the phrase "genetic takeover" to describe the victory of RNA over clay. He says, "In the end the supremacy of organic bio-materials is tied in with the question of scale. Organic machinery can be made much smaller. Such clever things become possible as sockets which can recognise, hold and manipulate other molecules. In any competition to do with molecular control the system with the smallest fingers will win."

At present there is no compelling reason to accept or to reject any of the three theories. Any of them, or none of them, could turn out to be right. We do not yet know how to design experiments that might decide between them. I happen to prefer the Oparin theory, not because I think it is necessarily right but because it is unfashionable. In recent years the attention of the experts has been concentrated upon the Eigen theory, and the Oparin theory has been neglected. The Oparin theory deserves a more careful analysis in the light of modern knowledge. In Chapter 3 I describe my own attempt to put the Oparin theory into a modern framework using the mathematical methods of Kimura.

There are numerous other theories that I have no time to discuss in detail. My physicist colleague Philip Anderson at Princeton proposed a theory that uses the concept of a spin glass in solid-state physics as a model of the transition to biological order (Anderson, 1983). I group Anderson's model together with Eigen's because both of them are primarily concerned with the replication of nucleic acids. A theory with more extensive experimental evidence to support it was proposed by Wächtershäuser. Wächtershäuser follows Cairns-Smith in conjecturing a stage of mineral evolution as a precursor to the evolution of organic life (Wächtershäuser, 1992, 1997). Where Cairns-Smith had clay as the precursor, Wächtershäuser has metal sulphides. Metal sulphides in general, and iron sulphides in particular, are good candidates for prebiotic chemistry and are known to be abundant in hydrothermal vents (Russell et al., 1994; Huber and Wächtershäuser, 1998). The Wächtershäuser theory fits well with the idea that life originated in a deep, hot environment. I confine my discussion here to the theories of Oparin, Eigen, and Cairns-Smith. Most of what I say about Eigen applies also to Anderson. Most of what I say about Cairns-Smith applies also to Wächtershäuser.

I find it illuminating to look at these theories in the light of the question that I raised in Chapter 1, whether the origin of life was a single or a double process, that is to say whether metabolism and replication originated together or separately. The Cairns-Smith theory is explicitly a double-origin theory. That to my mind is its main virtue. It has the first origin of life mainly concerned with the building of a protein metabolic apparatus, the clay particles adding to this apparatus a replicative element that may or may not be essential. The second origin of life, which Cairns-Smith calls "genetic takeover," is the replacement of the clay component by an efficient replicative apparatus made of nucleic acids. Cairns-Smith imagines the two origins of life to be separated by a long period of biochemical evolution so that the nucleic acid invasion occurs in cells already highly organized with protein enzymes and lipid membranes. The

Oparin theory and the Eigen theory were presented as single-origin theories. Each of them supposes the origin of life to have been a single process. Oparin places primary emphasis on metabolism and barely discusses replication. Eigen places primary emphasis on replication and imagines metabolism falling into place rapidly as soon as replication is established. I am suggesting that the Oparin and Eigen theories make more sense if they are put together and interpreted as the two halves of a double-origin theory. In this case, Oparin and Eigen may both be right. Oparin is describing the first origin of life and Eigen the second. With this interpretation, we combine the advantages of the two theories and eliminate their most serious weaknesses. Moreover, the combination of Oparin and Eigen into a double-origin theory is not very different from the theory of Cairns-Smith. Roughly speaking, Cairns-Smith equals Oparin plus Eigen plus a little bit of clay. All three theories may turn out to contain essential elements of the truth.

There is a possible analogy between the origin of life and the origin of elaborate body plans in higher organisms. Half an eon ago, after life had existed for about 3 eons, there was a sudden efflorescence of elaborate body plans. The efflorescence is known as the "Cambrian explosion" and produced in a geologically short time all the major body plans from which modern higher organisms evolved. Something must have happened shortly before the Cambrian epoch to make the genetic programming of elaborate body plans possible. What might have happened was the invention of "indirect development," the system by which an embryo sets aside a package of cells that are destined to grow into an adult, the body plan of the adult having no connection with the body plan of the embryo. The advantage of this system is that the embryo provides life support to the adult during the vulnerable stages of its early growth, whereas the adult is free to evolve elaborate and fine-tuned structures unconstrained by existing structures of the embryo. Three California paleontologists (Davidson et al., 1995) have collected evidence that the great majority of existing body plans arose from indirect development. This fact was overlooked until recently because the two best-known body plans, the chordate and the arthropod, are exceptions to the rule. The chordates and arthropods, the two most successful phyla of animals, probably began like the others with indirect development but later evolved a shortcut system of direct development with the adult body plan growing directly from the embryo. Almost all marine phyla still rely on indirect development to some extent. The sea urchin is a well-known example. The adult sea urchin grows out of a pouch of cells that have no function in the life of the embryo. After the adult form has completed its growth and can fend for itself, all the embryonic structures are jettisoned.

If the system of indirect development came first, it means that multicellular organisms evolved by a two-step process. The first step was the evolution of embryonic forms of limited complexity lacking the genetic machinery to program specialized structures. The second step was the evolution of adult forms with the modern armamentarium of genetic controls and with life support provided by the embryo. I am proposing that the early evolution of life followed the same two-step pattern as the evolution of higher organisms. First came the embryonic stage of life, cells with functioning metabolism but without any genetic apparatus, unable to evolve beyond a primitive level. Second came the adult stage, cells with genetic machinery allowing the evolution of far more finely tuned metabolic pathways and again with life support provided by the first stage while the second stage evolved.

To me, one of the most attractive features of the two-stage theory of the origin of life is that it shows life following the same pattern at three crucial periods of its history: first, the period of origins, when the two stages were metabolism and replication; second, the evolution of eucaryotic cells according to Margulis, when the two stages were parasitic invasion and symbiosis; and, third, the evolution of higher organisms, when the two stages were the embryo and the package of cells that grew into an adult. In each of the three revolutions, the first stage relied on crude and simple modes of inheritance, and the second stage jumped to new levels of sophistication in the translation of anatomical structure into genetic language.

CHAPTER THREE

A Toy Model

THE MEANING OF METABOLISM

This chapter describes my own attempt to understand the Oparin theory of the origin of life. The essential feature of the Oparin theory is that it has life beginning with metabolism rather than with precise replication. In this chapter I shall use the phrase "Oparin theory" to include both the original Oparin theory and the later versions proposed by Wächtershäuser and others. The Oparin theory as Oparin proposed it made no attempt to be quantitative. I am trying to place the theory within a framework of strict mathematics so that its consequences can be calculated. The essential difficulty arises because metabolism is a vague and ill-defined concept. There is no such difficulty with the concept of replication. Replication means exactly what it says. To replicate a molecule means to copy it, either exactly or with a stated margin of error. Starting from this well-defined concept, Manfred Eigen was able to formulate his theory of the origin of life, which is in fact a theory of the origin of replication, as a system of equations that can be solved with a computer. Eigen's equations describe the evolution with time of populations of molecules subject to nonlinear laws of replication. When we try to formulate in a similarly exact fashion the theory of Oparin, which is a theory of the origin of metabolism, we run immediately into the problem of defining what we mean by metabolism.

Doron Lancet has tackled this problem by studying computer models of the evolution of molecular populations, which he calls replicative-homeostatic early assemblies (RHEA). In these models, metabolism is defined in a general way as the evolution of a population in which some of the molecules catalyze the synthesis of others. He finds conditions under which populations can evolve to a high and self-sustaining level of catalytic organization. Many other computer studies of the evolution of metabolism have been published. The results are summarized in a recent review article (Segré and Lancet, 1999). My own model of molecular evolution is a very special case of a RHEA model. My model has an antique flavor because its behavior is simple enough to be calculated with pencil and paper rather than with computer simulations.

I reduce the Oparin theory to a mathematically precise form in two stages. The first stage is a formal description of molecular populations treating them like a classical dynamical system and making the dynamical equations precise but leaving the laws of interaction completely general. The general theory of molecular systems obtained in this way allows us to define what we mean by the origin of metabolism but does not allow us to predict under what conditions metabolism will occur. The second stage consists of the reduction of the general theory to a toy model by the assumption of a simple and arbitrary rule for the probability of molecular interactions. The entire intricate web of biochemical processes is replaced in the model by a couple of simple equations. The habit of constructing toy models of this sort is one to which theoretical physicists easily become addicted. When the real world is recalcitrant, we build ourselves toy models in which the equations are simple enough for us to solve. Sometimes the behavior of the toy model provides illuminating insight into the behavior of the real world. More often, the toy model remains what its name implies, a plaything for mathematically inclined physicists. In the present case, the toy model may have some connection with reality or it may not. Whether or not its premises are reasonable, at least its conclusions are definite. Given the premises from which it starts, it behaves as one would wish a primeval molecular population to behave, jumping with calculable probability between two states that differ by the presence or absence of metabolic organization. But before defining the toy model in detail, I go back to stage one and define the general theory of molecular evolution of which the toy model is a special case.

The general theory begins with an abstract multidimensional space of molecular populations. Each point of the space corresponds to a particular list of molecules that are supposed to be present at a particular moment in a particular population. The population is confined in a droplet, as Oparin imagined it. Small molecules that are free to diffuse from the surrounding medium into the droplet and out again are not counted. The population of molecules within the droplet can change from moment to moment, either by chemical reactions within the populations, or by reactions incorporating small molecules from the medium, or by reactions rejecting small molecules into the medium. The theory represents all these chemical reactions by a single matrix M of probabilities. Given a population in the state A, the next chemical reaction will take it to the state B with a certain probability, which is the matrix element of M between the states A and B. The population thus evolves in a stepwise and stochastic fashion over the space of possible states. The stepwise evolution can be described by a simple linear equation

$$P(k+1) = MP(k), \tag{3.1}$$

where $P(k)$ is the probability distribution of the population after k steps and $P(k+1)$ is the probability distribution after $(k+1)$ steps. This equation has the formal solution

$$P(k) = M^k P(0), \tag{3.2}$$

where $P(0)$ is the probability distribution of populations in an arbitrary initial state. We are interested in population distributions that persist during evolution over long periods. We call such distributions quasi-stationary. A quasi-stationary distribution Q satisfies an equation

$$MQ - Q = F, \tag{3.3}$$

where F is the flow vector describing the leakage of population out of the distribution Q, and F is small, of the order of m^{-1} for a distribution that persists through m chemical reactions.

There will in general be a finite number of quasi-stationary distributions of various degrees of longevity. Each quasi-stationary distribution has a basin of attraction, a region of configuration space surrounding it. All populations in any one basin are attracted to the same quasi-stationary distribution if they are allowed to evolve for a long time. The various basins of attraction cover the space without overlapping. Each quasi-stationary distribution Q has a leakage flow F, which defines the small probability of a population in the basin of attraction of Q escaping over the watershed into some neighboring basin.

Metabolism, unlike replication, is not a uniquely defined chemical process. Metabolism is a matter of degree. Some quasi-stationary distributions of molecules will metabolize more than others. Even a completely random and disorganized population of molecules will metabolize to some extent when placed in an environment providing a supply of fresh molecular components. The essential feature that we are looking for in a model of the origin of metabolism is a molecular system with two or more basins of attraction separated by high barriers. Each basin will then have a distinct quasi-stationary distribution within it. If we choose a random assortment of molecules and allow it to evolve stochastically in accordance with Eq. (3.1), it will generally end up in a particular quasi-stationary distribution that we may call the dead or disorganized state. Other quasi-stationary states that are separated from the disorganized state by high barriers are likely to be more structured, possibly with active biochemical cycles and higher rates of metabolism.

The decisive events in a theory of the origin of metabolism are the rare statistical jumps when a molecular population in one quasi-stationary state happens to undergo a succession of chemical reactions that push it up, against the gradient of probability, over a barrier and down into another quasi-stationary state. If the initial state is disorganized and the final state is organized, the jump may be considered to be a model for the origin of metabolism. In a complete theory of the origin of life it is likely that there would be several such jumps, each jump taking a population of molecules to a new quasi-stationary state, each quasi-stationary state having a more complex structure than the previous state. In my toy model I shall be concerned only with the first jump. In order to describe the first jump it is sufficient to have a model with two quasi-stationary states, one disordered and the other to some extent ordered or biochemically active. The barrier between the two states must be high enough to give the ordered state reasonable longevity but not so high as to make the jump from disorder to order practically impossible. We require the probability of the jump to be negligible for an individual droplet but still large enough so that jumps will occur occasionally in an assemblage of many droplets existing for a long time.

This is the end of stage one in the discussion of the origin of metabolism, the stage of abstract mathematical description without a specific model. The major obstacle we face in constructing a realistic theory is the absence of experimental information about possible metabolic cycles that are substantially simpler than the very complicated cycles we see in modern organisms. The primeval metabolic cycles must have been simpler than the modern ones, but we do not know what they were. We do not even have plausible candidates for the rudimentary enzymes that must have been the ultimate ancestors of modern enzymes. Even if we suppose that clay crystals or iron sulphide membranes helped the metabolic cycle to get started, we still have a wide choice of candidates for the organic components of the cycle. Lacking such plausible candidates, we cannot begin to calculate the probability that a population of molecules would make the jump into a self-sustaining metabolic cycle. But my intention is to construct a model that is specific enough for me to describe the jump in detail and calculate the probabilities. I therefore leave aside the definition of the primeval metabolic cycles as a job for the future. It is a job for chemists rather than for physicists.

I now proceed to stage two of my theory. I give a description of a model of the origin of metabolism without defining metabolism explicitly. The definition of metabolism is implicit in the description of the model. I leave it to the readers to judge whether the assumptions of the model are plausible. The chief virtue of the model is that its consequences are calculable. It makes quantitative and not altogether trivial statements about the molecules that might be capable of making the jump from disorder to metabolic activity.

DETAILS OF THE MODEL

The model is called a Toy Model of the Oparin Theory (Dyson, 1982). It is not intended to be realistic. It leaves out all the complicated details of real organic chemistry. Its purpose is to provide an idealized picture of molecular evolution that resembles in some qualitative fashion the Oparin picture of the origin of life. After I have described the toy model and deduced its consequences, I will return to the question, whether the behavior of the model has any relevance to the evolution of life in the real world. The model is an empty mathematical frame into which we may later try to fit more realistic descriptions of prebiotic evolution. My analysis of the model is an elementary exercise in population biology using equations borrowed from Fisher and Kimura. The equations are the same whether we are talking about a population of molecules in a droplet or a population of birds on an island (Kimura, 1970).

To define the model, I make a list of ten assumptions. The list begins with general statements but by the end the model is uniquely defined. This makes it easy to generalize the model by modifying only the more specific assumptions.

Assumption 1 (Oparin Theory). Cells came first, enzymes second, genes much later.

Assumption 2. A cell is a confined volume of fluid containing small organic molecules (monomers) in solution. The monomers are free to diffuse in and out of the cell. Inside the cell is a chemically active surface with a fixed number N of sites exposed to the fluid. The surface may be the boundary membrane of the cell, or it may be a separate structure in the cell's interior. Each site can adsorb a monomer onto the surface, and thus monomers are continually exchanged between the surface and the fluid. Monomers adsorbed onto neighboring sites will link together to form polymers. To make the model more concrete, we may imagine the surface to be a clay crystal according to Cairns-Smith (1982) or an iron-sulphide membrane according to Wächtershäuser (1992) and Russell et al. (1994). We may imagine the monomers to be primitive versions of the amino acids that polymerize to make modern enzymes.

Assumption 3. Cells do not interact with one another. There is no Darwinian selection. Evolution of the population of molecules within a cell proceeds by random drift. Darwinian selection only begins after the model ends, as we shall see later.

Assumption 4. Changes of population occur by discrete steps, each step consisting of an adsorption or desorption of a single monomer at a single site of the surface. This assumption is unnecessarily restrictive and is imposed only for the sake of simplicity. At the cost of some complication of the mathematics, we could include a more realistic variety of chemical processes; for example, the linking of monomers into polymer chains in solution, or the adsorption and desorption of polymer chains at the surface.

Assumption 5. Each of the N sites on the surface adsorbs and desorbs monomers with equal probability. This assumption is also unrealistic and is made to keep the calculations simple.

Assumption 6. The monomers bound to the surface can be divided into two classes, active and inactive. This assumption appears to be uncontroversial, but it actually contains the essential simplification that makes the model mathematically tractable. It means that we are replacing the enormous multidimensional space of molecular configurations by a single variable, taking only two values, 1 for "active" and 0 for "inactive."

Assumption 7. The active monomers are those that happen to be of the right species at the right sites, where they and their neighbors make a polymer that can act as an enzyme. To act as an enzyme means to catalyze the adsorption of other monomers in a selective manner so that monomers of the correct species are chosen preferentially to be adsorbed at other sites where they can be active.

Assumption 7 is the place where the notion of metabolism sneakily enters the model without being defined. The definition of the word "active" in Assumption 7 is circular. An active monomer is one that helps other monomers to be active. Metabolism, as defined by the model, means the cyclic shuffling around of monomer units, and the population as a whole is metabolically active if the cyclic shuffling maintains the active monomers at a self-sustaining high level.

Assumption 8. The monomers belong to $(n+1)$ chemical species, all present in the fluid with equal abundance. At each site, only one species is active, and the remaining n species are inactive. Each empty site will adsorb each inactive species of monomer with equal probability p per unit time. The total rate np of adsorption of inactive monomers is the same at every site. The monomer at each filled site will be desorbed with probability qp per unit time independent of whether it is active or inactive. Here q is a constant, depending on the temperature and on the strength of the attraction between the surface and the monomers. Each empty site will adsorb its active species of monomer with probability $\psi(x)p$ per unit time, where x is the fraction of all sites in the cell already occupied by active monomers. The function $\psi(x)$ represents the efficiency of the existing population of active monomers in accelerating the adsorption of a new active monomer. Each event of adsorption or desorption can be regarded as an act of reproduction changing a parent population into a daughter population. The assumption that $\psi(x)$ depends on x means that the activity of monomers in the parent population is to some extent inherited by a newly adsorbed monomer in the daughter population. The form of $\psi(x)$ expresses the law of inheritance from parent to daughter. The numerical value of $\psi(x)$ will be determined by the details of the chemistry of the catalysts.

Assumption 8 is a drastic approximation. It replaces the average of the efficiencies of a population of catalysts by the efficiency of an average catalyst. I call it the "mean field approximation" because it is analogous to the approximation made in the Curie–Weiss mean-field model of a ferromagnet. In physics, we know that the mean-field approximation gives a good qualitative account of the behavior of a ferromagnet. In population biology, similar approximations have been made by Kimura. The effect of the mean-field approximation is to reduce the multidimensional random walk of molecular populations to a one-dimensional random walk of the single parameter x. Both in physics and in population biology, the mean-field approximation may be described as pessimistic. It underestimates the effectiveness of local groupings of molecules in forming an ordered state. The mean-field approximation generally predicts a lower degree of order than is found in an exact theory.

If we imagine a cell with the population of monomers on the surface in a steady state, Assumption 8 will have the following consequences. The fractions (x, w, z) of active, inactive, and empty sites will in a steady state be proportional to $(\psi(x), n, q)$. The fractions (x, w, z) must add up to unity, and therefore

$$x = \phi(x) = \psi(x)/(\psi(x) + a) = (1 + a(\psi(x))^{-1})^{-1}, \tag{3.4}$$

with

$$a = n + q. \tag{3.5}$$

In a state that is not steady, the ratio $(\phi(x)/x)$ is roughly the ratio of the rate of increase of active sites by adsorption to the rate of decrease of active sites by desorption. In a steady state this ratio must be unity.

Assumption 9 (Fig. 3). The curve $y = \phi(x)$ is S-shaped, crossing the line $y = x$ at three points, $x = \alpha, \beta, \gamma$, between zero and one. This assumption is again borrowed from the Curie–Weiss model of a ferromagnet. It means that the population of molecules has three possible equilibrium states. An equilibrium state occurs whenever $\phi(x) = x$, when the daughter population inherits the same average activity x as the parent population. The equilibrium is stable if the slope of the curve $y = \phi(x)$ is less than unity and is unstable if the slope is greater than unity. Consider, for example, the lowest equilibrium state $x = \alpha$. I call it the disordered state because it has the smallest average activity. Because the slope at α is less than unity, the equilibrium is stable. If a parent population has its average activity x a little above α, the daughter population will tend to slide back down toward α. If the parent population has x a little below α, the daughter population will tend to slide up toward α. The same thing happens at the upper equilibrium state $x = \gamma$. The upper state is also stable because the slope at γ is less than unity. I call it the

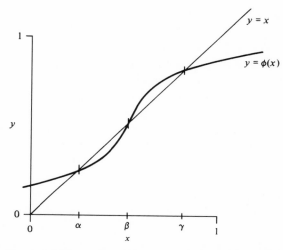

Figure 3   The S-shaped curve $y = \phi(x)$. The curve crosses the line $y = x$ at three points that represent possible equilibrium populations. The upper and lower equilibria are stable, and the middle equilibrium is unstable.

ordered state because it has the largest catalytic activity. A population with activity x close to γ will move closer to γ as it evolves. But the middle equilibrium point $x = \beta$ is unstable because the slope at β is greater than unity. If a population has x slightly larger than β, it will evolve away from β toward the ordered state $x = \gamma$, and if it has x slightly smaller than β it will slide away from β down to the disordered state $x = \alpha$. The equilibrium at $x = \beta$ is an unstable saddle point.

Going back to the abstract mathematical language of stage one of the theory, we may say that the lower and upper equilibrium states of the model are two quasi-stationary states occupying two separate basins of attraction. The barrier separating the two basins is a mountain pass with the unstable saddle point equilibrium state at the top.

We have here a situation analogous to the distinction between life and death in biological systems. I call the ordered state of a cell "alive" because it has most of the molecules working together in a collaborative fashion to maintain the catalytic cycles that keep them active. I call the disordered state "dead" because it has the molecules uncoordinated and mostly inactive. A population, either in the dead or in the alive state, will generally stay there for a long time, making only small random fluctuations around the stable equilibrium. However, the population of molecules in a cell is finite, and there is always the possibility of a large statistical fluctuation that takes the whole population together over the saddle point from one stable equilibrium to the other. When a "live" cell makes the big statistical jump over the saddle point to the lower state, we call the jump "death." When a "dead" cell makes the jump up over the saddle point to the upper state, we call the jump "origin of life." Once the function $\phi(x)$ and the size N of the population in the cell are given, the probabilities of "death" and of the "origin of life" can easily be calculated. We have only to solve a linear difference equation with the appropriate boundary conditions to represent an ensemble of populations of molecules diffusing over the saddle point from one side or the other.

The probability of the transition over the saddle point is given precisely by the flow term F in Eq. (3.3) of the general mathematical description. It is perhaps misleading to use the word "jump" to describe the transition. The word "diffusion" describes it better. The population of molecules climbs up the barrier by a succession of numerous small steps rather than by one big step. The population will probably take nine steps backward for every ten steps forward as it inches its way uphill. Fortunately, the net flow F can easily be calculated without following the details of the random walks by which the flow is achieved.

Assumption 10. Here we make a definite choice for the function $\psi(x)$, basing the choice on a simple thermodynamic argument. It will turn out happily that the function $\phi(x)$ derived from thermodynamics has the desired S-shaped form to produce the three equilibrium states required by Assumption 9.

We assume that every catalyst in the cell works by producing a difference between the activation energies required for placing an active or inactive monomer into an empty site. If the catalyst molecule is perfect, with all its monomer units active, then the difference in activation energies will be a certain quantity U which we assume to be the same for all perfect catalysts. If a catalyst is imperfect, in a cell with a fraction x of all sites active, we assume that it produces a difference xU in the activation energies for correct and incorrect adsorptions. We are here again making a mean-field approximation, assuming that the average effect of a collection of catalysts with various degrees of imperfection is equal to the effect of a single catalyst with its discrimination reduced in proportion to the average activity x of the whole population. This is another approximation that could be avoided in a more exact calculation.

The mean-field approximation implies that the probability of a correct adsorption is increased over the probability of an incorrect adsorption by the factor

$$\psi(x) = b^x, \qquad b = \exp(U/kT), \tag{3.6}$$

so that b is the discrimination factor of a perfect catalyst at absolute temperature T, and k is Boltzmann's constant. This choice of $\psi(x)$ gives for the function $\phi(x)$ according to Eq. (3.4),

$$\phi(x) = (1 + ab^{-x})^{-1}, \tag{3.7}$$

which is the same S-shaped function that appears in the mean-field model of a simple ferromagnet.

The formula (3.7) for $\phi(x)$ completes the definition of the model. The model is uniquely defined once the three parameters N, a, and b are chosen. The three parameters summarize in a simple fashion the chemical raw material with which the model is working. N defines the size of the molecular population, a defines the chemical diversity of the monomer units, and b is the quality factor defining the degree of discrimination of the catalysts. Strictly speaking, a is the sum of n and q, where $(n + 1)$ is the number of monomer species and q is proportional to the fraction of empty sites on the surface. So far as the model is concerned, the empty sites behave like an additional species of inactive monomer. The empty sites have statistical weight q compared with unity for the other inactive species. We assume that the monomers in the cell are abundant enough and the surface attractive enough, so that the fraction of empty sites is small and q is less than unity. Then the difference between n and a is insignificant.

We now have a definite three-parameter model to work with. We still have to calculate its consequences and to examine whether it shows interesting behavior for any values of N, a, and b that are consistent with the facts of organic chemistry. "Interesting behavior" here means the occurrence with reasonable probability of a jump from the disordered state to the ordered state. We shall find that interesting behavior occurs for values of a and b lying in a narrow range. This narrow range is determined only by the mathematical properties of the exponential function and is independent of the details of the chemistry. The model therefore makes a definite statement about the stuff out of which the first living cells were made. If the model has anything to do with reality, then the primeval cells were composed of molecules having values of a and b within the calculated range.

It turns out that the preferred ranges of values of the three parameters are as follows:

a, from 8 to 10; $\qquad\qquad\qquad\qquad\qquad\qquad$ (3.8)

b, from 60 to 100; $\qquad\qquad\qquad\qquad\qquad\qquad$ (3.9)

N, from 2000 to 20000. $\qquad\qquad\qquad\qquad\qquad$ (3.10)

These ranges also happen to be reasonable from the point of view of chemistry. The preferred range of values for a (3.8) says that the number of species of monomer should be in the range from eight to eleven. In modern proteins we have twenty species of amino acids. It is reasonable to imagine that about ten of them would provide enough diversity of catalytic function to get life started. On the other hand, the model definitely fails to work with a between 3 and 4, which would be the required range for a if life had begun with four species of nucleotides polymerizing to make RNA. Nucleotides alone do not provide enough chemical diversity to allow a transition from disorder to order in this model. The quantitative predictions of the model are thus consistent with the Oparin theory from which we started. The model prefers peptides to nucleic acids as the stuff from which life arose. In this respect the model differs from the Eigen hypercycle model of the origin of life. In the Eigen model the monomers are required only to carry information for the purpose of accurate replication. Four species of monomers, or even two species, are enough for replication. It is only when we require the monomers to function as a metabolic system that we need the greater diversity that our model dictates.

The range (3.9) from sixty to one hundred is also reasonable for the discrimination factor of primitive enzymes. A modern polymerase enzyme typically has a discrimination factor of 5000 or 10000. The modern enzyme is a highly specialized structure perfected by three thousand million years of fine-tuning. It is not to be expected that the original enzymes would have come close to modern standards of performance. On the other hand, simple inorganic catalysts frequently achieve discrimination factors of fifty. It is plausible that a simple peptide catalyst with an active site containing four or five amino acids would have a discrimination factor in the range preferred by the model, from sixty to one hundred.

The size (3.10) of the population in the primitive cell is also plausible. A population of several thousand monomers linked into a few hundred polymers would give a sufficient variety of structures to allow interesting catalytic cycles to exist. A value of N of the order of 10000 is large enough to display the chemical complexity characteristic of life and still small enough to allow the statistical jump from disorder to order to occur on rare occasions with probabilities that are not impossibly small.

The basic reason for the success of the model is its ability to tolerate high error rates. The model overcomes the error catastrophe by abandoning exact replication. It neither needs nor achieves precise control of its molecular structures. It is this lack of precision that allows a population of 10000 monomers to jump into an ordered state without invoking a miracle. In a model of the origin of life that assumes exact replication from the beginning, with a low tolerance of errors, a jump of a population of N monomers from disorder to order will occur with probability of the order of $(1 + n)^{-N}$. If we exclude miracles, a replicating system can arise spontaneously only with N of the order of one hundred or less. In contrast, our nonreplicating model can make the transition to order with a population that is a hundred times larger. The error rate in the ordered state of our model is typically between 20 and 30 percent when the parameters a and b are in the ranges (3.8) and (3.9), respectively. An error rate of 25 percent means that three out of four of the monomers in each polymer are correctly placed. A catalyst with five monomers in its active site has one chance out of four of being completely functional. Such a level of performance is tolerable for a nonreplicating system but would be totally unacceptable in a replicating system. The ability to function with a 25 percent error rate is the decisive factor that makes the ordered state in our model statistically accessible with populations large enough to be biologically interesting.

CONSEQUENCES OF THE MODEL

The equations describing a stationary state of the model are very simple. In a quasi-equilibrium state there will be a certain probability P_j for finding j active monomers in the population. The flow-rate F in a steady state will be independent of j. The equation (3.3) defining a quasi-equilibrium state then becomes

$$(q/a)\psi(j/N)(N - j)P_j - q(j + 1)P_{j+1} = F. \tag{3.11}$$

It is not difficult to solve Eq. (3.11) exactly and deduce the flow F that determines the rate of transitions from one quasi-equilibrium state to another. But we can estimate the rate of transitions more easily, with sufficient accuracy for our purposes, by solving Eq. (3.11) with $F = 0$. When $F = 0$, the solution P_j of Eq. (3.11) describes the unique final state of the model when it is allowed to run for an infinite time. In the final state, all quasi-equilibrium states are populated so that the net flows between them are zero. The solution of Eq. (3.11) is then, with $\psi(x)$ given by Eq. (3.5), and after making some unimportant approximations

$$P_j = K \exp(-NV(j/N)), \tag{3.12}$$

with a constant coefficient K and a potential $V(x)$ given by

$$V(x) = x \log x + (1 - x) \log(1 - x) + x \log a - (1/2)x^2 \log b.$$

$$(3.13)$$

The quasi-equilibrium states of the model correspond precisely to the maxima and minima of the potential V. The stable quasi-equilibrium states $(x = \alpha, \gamma)$ are minima of V; the unstable state $(x = \beta)$ is a maximum.

In this way we find an approximate formula for F, which is equivalent to the simple formula

$$T = \tau \exp(\Delta N) \qquad\qquad (3.14)$$

for the average time T required for a cell to make the transition from disorder to order. Here τ is the average time interval between desorptions of a monomer at each site, N is the total number of sites, and Δ is given by

$$\Delta = V(\beta) - V(\alpha). \qquad\qquad (3.15)$$

This Δ is the height of the potential barrier that the population has to climb to escape from the disordered state $(x = \alpha)$ over the saddle point $(x = \beta)$ to the ordered state. The rate of the transition from disorder to order does not depend on $V(\gamma)$. Once the population crosses the saddle point, the time required to slide down to the ordered state is negligible.

If Δ were of the order of unity, then the exponential in Eq. (3.14) would be impossibly large for N greater than one hundred. We would then be in the situation characteristic of error-intolerant systems, for which the transition to order is astronomically improbable for large N. However, when the parameters a and b are in the ranges (3.8) and (3.9), respectively, which correspond to models with high error tolerance, it turns out that Δ is not of the order of unity but lies in the range from 0.001 to 0.015. For values of a and b in these ranges, the potential $V(x)$ is almost flat, and its values at the three stationary points are almost equal. These are the features of the model that make a transition to order possible with populations as large as 20000. Although Eq. (3.14) is still an exponentially increasing function of N, it increases much more slowly than one would naively expect.

According to Eq. (3.14), there is a critical population-size N_c such that populations N of the order of N_c or smaller will make the disorder-to-order transition with reasonable probability, whereas populations much greater than N_c will not. I choose to define N_c by

$$N_c = 30/\Delta, \tag{3.16}$$

so that the exponential factor in Eq. (3.14) is

$$e^{30} \approx 10^{13} \text{ for } N = N_c. \tag{3.17}$$

The coefficient thirty in Eq. (3.16) is chosen arbitrarily. We do not know how many droplets might have existed in environments suitable for the origin of life, nor how long such environments lasted, nor how frequently their molecular constituents reacted. The choice in Eq. (3.16) means that we could expect one transition to the ordered state to occur in a thousand reaction times among a collection of 10^{10} droplets, each containing N_c monomers. It is not absurd to imagine that 10^{10} droplets may have existed for a suitably long time in an appropriate environment. On the other hand, if we were to consider droplets with molecular populations three times larger, that is to say with $N = 3N_c$, then the exponential factor in Eq. (3.14) would be 10^{39}, and it is inconceivable that enough droplets could have existed to give a reasonable probability of a transition. The critical population N_c thus defines the upper limit of N for which transition can occur with a margin of uncertainty that is less than a factor of three. The critical population sizes given by Eq. (3.16) range from 2000 to 20000 when the parameters a and b lie in the ranges 8–10 and 60–100, respectively.

The properties of our model can be conveniently represented in a two-dimensional diagram (Fig. 4) with the parameter a horizontal and the parameter b vertical. Each point on the diagram corresponds to a particular choice of a and b. Models that satisfy the triple-crossing condition (Assumption 9) and possess disordered and

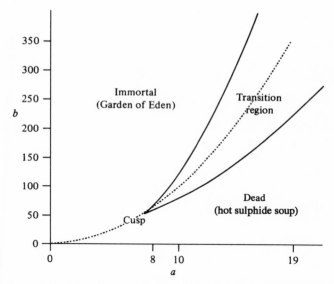

Figure 4 Phase diagram in which each point represents a possible chemical composition of a molecular population. Plotted horizontally is a, approximately equal to the number of species of monomers. Plotted vertically is b, the quality factor of polymer catalysts. The transition region represents populations that possess ordered and disordered equilibrium states. The "Dead" region has no ordered state, and the "Immortal" region no disordered state. The dotted curve represents populations with $b = a^2$ for which there is a symmetry between ordered and disordered states.

ordered states occupy the central region of the diagram extending up and to the right from the cusp. The cusp at

$$a = e^2 = 7.4, \qquad b = e^4 = 54.6, \tag{3.18}$$

marks the lower bound of the values of a and b for which a disorder–order transition can occur. The critical population-size N_c is large near the cusp and decreases rapidly as a and b increase. The biologically interesting models are to be found in the central region close to the cusp. These are the models that have high error rates and can make the disorder–order transition with large populations.

To illustrate the behavior of the model in the interesting region near the cusp, I pick out one particular case that has the advantage of being easy to calculate exactly. This is the case

$$a = 8, \qquad b = 64. \tag{3.19}$$

This has the three equilibrium states

$$\alpha = (1/3), \qquad \beta = (1/2), \qquad \gamma = (2/3). \tag{3.20}$$

The error rate in the ordered state is exactly one-third. The value of Δ for this model is given by Eqs. (3.13) and (3.15),

$$\Delta = \log 3 - (19/12) \log 2 = 0.001129, \tag{3.21}$$

which gives a satisfactorily large critical population size

$$N_c = 26566. \tag{3.22}$$

Christopher Longuet-Higgins, who happens to be a musician as well as a chemist, pointed out that the quantity Δ appearing in Eq. (3.21) is well known to musicians as the fractional difference in pitch between a perfect fifth and an equitempered fifth. On a logarithmic scale of pitch, a perfect fifth is $(\log 3 - \log 2)$ and an equitempered fifth is seven semitones or $(7/12) \log 2$. The smallness of the difference is the reason why the equitempered scale works as well as it does. The smallness of Δ is also the reason this model of the origin of life works as well as it does. Old Pythagoras would be pleased if he could see this example justifying his doctrine of a universal harmony that embraces number, music, and science. After this digression into Pythagorean mysticism I return to the general properties of the model shown in Fig. 4.

The region below and to the right of the central strip represents models that have only a disordered state and no ordered state. These models have a too large (too much chemical diversity) and b too small (too weak a catalytic activity) to produce an ordered state. Droplets in this region are dead and cannot come to life. I call the region "hot sulphide soup" because this phrase has been used to describe the composition of deep hydrothermal vents where life might have originated according to Wächtershäuser (1992) and Russell et al. (1994). The phrase "cold chicken soup," which was used in earlier times to describe the prebiotic environment in which life arose, is no longer appropriate, since many of the most archaic organisms have been found to be thermophilic. The region above and to the left of the central strip represents models that have only an ordered state and no disordered state. These models have a too small (too little chemical diversity) and b too large (too strong a catalytic activity) to produce a disordered state. Droplets in this region are frozen into the ordered state and cannot die. I call the region "Garden of Eden" because this phrase has been used to describe an alternative theory of the origin of life. It is possible to imagine cells evolving by random accretion of molecular components so that they drift into the central transition region either from the hot sulphide soup or from the Garden of Eden. Once they reach the central region, they are capable of life and death, and the evolution of biological complexity can begin.

Why do we need disordered states? Because life is an ordered state, why should not life have remained for ever in the "Garden of Eden" region where only ordered states exist? The model does not supply an answer to these questions, because it only describes random drift of populations without Darwinian selection. So far as the model is concerned, life could have begun and remained permanently frozen in the ordered state. But in fact, Darwinian selection was essential if life was to evolve beyond the primitive stage described by the model. After the model ends, as soon as the droplets described by the model begin to exhaust the available resources and to compete with one another for survival, Darwinian evolution begins. Darwinian selection requires death, and death is the transition from order to disorder. Life had to invent death to evolve. Droplets frozen into the ordered state could not experiment and diversify. The droplets that stayed in the transition region, constantly dying and being reborn with new combinations of chemicals in new catalytic cycles, had the best chance of adapting fast to changing environments.

As molecular populations evolve over long periods, it is likely that a and b will increase. An increase in a will result from incorporation of new varieties of monomer into the metabolic cycle. An increase in b will result from the buildup of more complicated

polymer molecules able to catalyze reactions with greater specificity. Molecular populations with larger a and b will metabolize
more efficiently and are likely to prevail in the Darwinian struggle
for existence. After our model ends, evolution can proceed gradually upward and to the right while remaining within the critical
strip of Fig. 4. It is only necessary for a and b to keep roughly in
step so that b is approximately equal to the square of a. The curve
$b = a^2$ remains within the critical strip all the way from the cusp
to infinity. Of course, as evolution proceeds further and further up
the strip, the simple assumptions of the model will represent the
behavior of real organisms more and more inadequately.

The model allows us to give a precise definition of the quantity
of information contained in a population in the ordered state. The
information is not embodied in genes that can be replicated. It is
embodied in the metabolic cycles that are reproduced statistically as
the population maintains itself. According to the abstract definition
of information, the probability of a message arising by pure chance
is given by

$$P = 2^{-I}, \tag{3.23}$$

where I is the number of bits of information in the message. To apply
this definition to the model, we imagine the ordered and disordered
states to be coexisting in the proportions dictated by pure chance.
The probability of the ordered state is then, according to Eq. (3.12)

$$P = \exp[N(V(\alpha) - V(\gamma))], \tag{3.24}$$

and the information contained in it is by Eq. (3.23)

$$I = N(V(\gamma) - V(\alpha))/(\log 2). \tag{3.25}$$

This gives a minimum estimate of the information content of the
ordered state. The cell may also contain additional information that
does not contribute to making it statistically improbable.

At the moment of the first jump from disorder to order, at the
first origin of metabolism, the information content of the ordered
state in our model is quite small. It is of the order of ΔN, which
we assumed to be about 30, and thus the jump to the ordered state

can happen with reasonable probability in a finite time. Roughly speaking, Eq. (3.25) then gives $30/\log 2 = 43$ bits of information. This is a very small quantity of information. But the quantity of information carried by the ordered state increases rapidly as the cell evolves along the path of increasing N and increasing a and b, that is to say, increasing population and increasing variety and precision of enzymes. When both a and b are large, the information given by Eqs. (3.13) and (3.25) becomes

$$I = (1/2)N(\log(b/a^2)/\log 2). \tag{3.26}$$

In the symmetric version of the model with $b = a^2$, the ordered and disordered equilibrium states are equally probable. The formula (3.26) gives zero because the ordered state occurs with probability one half if you wait long enough. In a modern, well-organized cell, typical values of the parameters are $a = 20$ and $b = 10^4$, Eq. (3.26) gives roughly

$$I = 2N, \tag{3.27}$$

about two bits of information per site in the active catalytic molecules. This quantity of information happens to be equal to the information contained in a gene with N nucleotides. Thus the metabolic apparatus in our model carries about as much information as a replicative apparatus with the same number of active monomers.

One striking feature of the model that is absent in modern organisms is the symmetry between life and death. In the model, the curve

$$y = \phi(x) = (1 + ab^{-x})^{-1} \tag{3.28}$$

is invariant under the transformation

$$x \to 1 - x, \qquad y \to 1 - y, \qquad a \to (b/a). \tag{3.29}$$

In particular, the model with $b = a^2$ has complete symmetry about the unstable saddle point at $x = y = 1/2$. The ordered state and the disordered state are mirror images of each other. The probability of a transition from disorder to order is exactly equal to the probability of a transition from order to disorder. In the symmetrical model

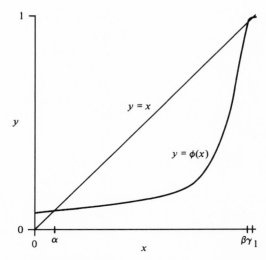

Figure 5   The S-shaped curve of Fig. 3 as it appears in a modern cell with the upper and middle equilibrium states pushed over far to the right so that the curve is no longer symmetrical.

with $b = a^2$, death and resurrection occur with equal frequency. The origin of life is as commonplace an event as death.

How did it happen that, as life evolved, death continued to be commonplace while resurrection became rare? What happened was that the catalytic processes in the cell became increasingly fine-tuned and increasingly intolerant of error. The curve $y = \phi(x)$ remained S-shaped but became more and more asymmetrical as time went on.   The shape of the curve in a modern cell is shown in Fig. 5. This should be contrasted with the symmetrical curve shown in Fig. 3 for our hypothetical primitive cell. In the primitive cell the three equilibrium states might have been

$$\alpha = 0.2, \qquad \beta = 0.5, \qquad \gamma = 0.8 \tag{3.30}$$

with an error-rate of 20 percent in the ordered state. In the modern cell the curve is pushed over far to the right, and for the equilibrium states the values are typically

$$\alpha = 0.05, \qquad \beta = 0.999, \qquad \gamma = 0.9999. \tag{3.31}$$

This value of the ordered state γ means that the error rate in the metabolic apparatus of a modern cell is about 10^{-4}. The position of the saddle point β means that an environmental insult such as a dose of X-rays that increases the error-rate to 10^{-3} will disrupt the fine-tuned apparatus and cause the cell to die. Death is easy, and resurrection is difficult because the saddle point has moved so close to the ordered state and so far from the disordered state. For life to originate spontaneously it was essential to have an ordered state with a high error rate, but when life was once established the whole course of evolution was toward more specialized structures with lower tolerance of errors.

I have said enough, or perhaps too much, about the properties and the consequences of the model. In talking about the model I have fallen into a trap. I begin to talk about it as if it were historic truth. It is of course nothing of the kind. It is not a description of events as they really happened. It is only a toy model, a simple abstract picture that will rapidly be superseded by better models incorporating some of the chemical details that I have ignored.

CHAPTER FOUR

Open Questions

WHY IS LIFE SO COMPLICATED?

It is now time to sum up what we may have learned from the first three chapters. Chapter 1 describes the historical development of ideas leading up to the question that I consider to be fundamental to all investigation of the origin of life: Is the origin of life the same thing as the origin of replication? I give some reasons why I am inclined to answer no to this question, to give a tentative preference to the hypothesis that metabolism and replication had separate origins. Chapter 2 gives a sketchy account of some of the classic experiments and some of the classic theories concerning the origin of life. I observe that the experiments since the time of Max Delbrück have been spectacularly successful in elucidating the structure and function of the apparatus of replication and much less successful in giving us a deep understanding of metabolism. Although the experiments of Cech and others (Cech, 1993; Wright and Joyce, 1997) on ribozymes have demonstrated that RNA can function as an enzyme, it acts as an enzyme only within a limited domain. An early review article (Cech and Bass, 1986) said, "It appears to be the limited versatility of RNA catalysts, rather than any deficit in catalytic efficiency or accuracy, that is responsible for the relatively restricted occurrence of RNA as a biological catalyst." It remains true today that the experiments investigating the action of ribozymes are concerned with the fine-tuning of the genetic apparatus, not with the metabolism of the cell. The one-sided success of the experiments has resulted in a corresponding bias of theories. The most popular theories of the origin of life are the theories of Manfred Eigen, which concentrate almost exclusively on replication as the phenomenon to be explained. Chapter 3 describes my own attempt to build a model of the origin of life with a bias opposite to Eigen's, assuming as a working hypothesis that primitive life consisted of purely metabolic machinery without replication.

This last chapter is concerned with the open questions raised by the model and with more general questions concerning possible experimental approaches to the origin of metabolism. But all these questions are subsidiary to another question, Why is life so complicated? This is perhaps not a well-posed scientific question. It might be interpreted as merely the lament of an elderly scientist recalling the lost simplicity of youth. Or it might be interpreted as an ineffectual protest against the intractability of the human condition in the modern world. But I mean the question to refer specifically to cellular structure. The essential characteristic of living cells is homeostasis, the ability to maintain a steady and more-or-less constant chemical balance in a changing environment. Homeostasis is the machinery of chemical controls and feedback cycles that make sure that each molecular species in a cell is produced in the right proportion, not too much and not too little. Without homeostasis, there can be no ordered metabolism and no quasi-stationary equilibrium deserving the name of life. The question Why is life so complicated? means, in this context, Given that a population of molecules is able to maintain itself in homeostatic equilibrium at a steady level of metabolism, how many different molecular species must the population contain?

The biological evidence puts rather firm limits to the number of kinds of molecule needed to make a homeostatic system, at least so long as we are talking about homeostatic systems of the modern type. There is a large number of different varieties of bacteria, and most of them contain a few thousand molecular species, if one judges the number by the few million base pairs in their DNA. It seems that under modern conditions homeostatic systems work efficiently with a few thousand components and work less efficiently with fewer. If a bacterium could dispense with half of its molecular components and still metabolize efficiently, there would be a great selective advantage in doing so. From the fact that bacteria have generally refused to shrink below a certain level of complexity, we may deduce that this level is in some sense an irreducible minimum.

If modern cells require a few thousand types of molecule for stable homeostasis, what does this tell us about primitive cells? Strictly speaking, it tells us nothing. Without the modern apparatus of genes and repressors, the ancient mechanisms of homeostasis must have been very different. The ancient mechanisms might have been either simpler or more complicated. Still it is a reasonable hypothesis that the ancient mechanisms were simpler. There remains the question, How simple could they have been? This question must be answered before we can build credible theories of the origin of life. It can be answered only by experiment.

In the toy model that I discuss in Chapter 3, I deduced from the arithmetic of the model that the population of a cell making the transition from disorder to order should have been between 2000 and 20000 monomers combined into a few hundred species of polymers. I claim that this number, a few hundred, is plausible for the number of species of polymer molecule required for a primitive homeostatic system. That claim is, of course, based on nothing but guesswork. We know that a few thousand species of molecule are sufficient for a modern cell. It seems unlikely that anything resembling biochemical homeostasis could be maintained with a few tens of species. And so we guess quite arbitrarily, guided only by our familiarity with the decimal system of counting, that a few hundred kinds of molecule are the right number for the origin of homeostasis. Whether a few hundred molecular species are really either necessary or sufficient for homeostasis we do not know.

It is of interest in this connection to see how an experimental approach was successfully applied to answer the corresponding question concerning the origin of replication. What is the smallest molecular population that is able to constitute a self-replicating system? This question was answered by two classic experiments, one done by Sol Spiegelman (Spiegelman, 1967), the other by Manfred Eigen and his colleagues (Eigen et al., 1981). I describe Eigen's experiment in Chapter 1. Spiegelman's experiment began with a living Q_β virus, a creature able to survive and ensure its own replication in nature by means of the information coded in a single RNA molecule composed of 4500 nucleotides. The virus is normally replicated inside a host cell using a replicase enzyme that the viral RNA causes the host's ribosomes to manufacture. The viral RNA also causes the host to manufacture a coat protein and various other components that are required for the complete viral life cycle. Now, Spiegelman proceeded to debauch the virus by stripping off its protein coat and providing it with replicase enzyme in a test tube so that it could replicate without going to the trouble of invading a cell and completing its normal parasitic life cycle. The test tube also contained an ample provision of free nucleotide monomers with a continuous-flow arrangement to keep the virus from exhausting the supply. The results were spectacular. The viral RNA continued for a while to be replicated accurately with the help of the replicase enzyme. But soon a mutant RNA appeared, having lost some of the genes that were no longer required for its survival. The mutant, having fewer than 4500 nucleotides, was replicated more quickly than the original virus and soon displaced it in the Darwinian struggle for existence. Then another still shorter mutant appeared to displace the first, and so it went on. The virus no longer needed to carry the genes for replicase and coat protein to survive. On the contrary, it could only survive by getting rid of all superfluous baggage. The requirement for survival was to be as simple and as small as possible. The virus finally degenerated into a little piece of RNA with only 220 nucleotides containing the recognition site for the replicase enzyme and not much else. The final state of the virus was called by geneticists the "Spiegelman monster." It provides a good object lesson demonstrating what happens to you when life is made too easy. The little monsters would continue forever to replicate at high speed in the artificial environment of Spiegelman's test tube but could never hope to survive anywhere else.

The experiment of Manfred Eigen was the opposite of Spiegelman's experiment. Both experiments used a test tube containing replicase enzyme and free nucleotides. Spiegelman put into this soup a living virus. Eigen put in nothing. Spiegelman was studying the evolution of replication from the top down, Eigen from the bottom up. Eigen's experiment produced a self-generated population of RNA molecules replicating with the help of the replicase enzyme, just like the Spiegelman monsters. The Eigen replicator and the Spiegelman monster were not identical, but they were first cousins. Eigen's replicators, after they had evolved to a steady state, contained about 120 nucleotides each compared with the 220 in a Spiegelman monster. The difference between 120 and 220 nucleotides is a small gap between a molecule that grew from nothing and a molecule that was once alive.

The experiments of Spiegelman and Eigen together give a clear answer to the question, What is the minimum population size required for a replicating system? The answer is a single RNA molecule with one or two hundred nucleotides. This answer shows in a nutshell how simple the phenomenon of replication is compared with the phenomenon of homeostasis. I am conjecturing that the minimum population size required for homeostasis would be about a hundred times larger, namely, a few hundred molecules containing ten or twenty thousand monomer units. And more important, I am suggesting that the most promising road to an understanding of the origin of life would be to do experiments like the Spiegelman and Eigen experiments but this time concerned with homeostasis rather than with replication.

How could such experiments be done? I am acutely aware that it is much easier to suggest experiments than to do them. What is required first of all is to find the working materials that make experiments possible, the equivalent for a homeostatic system of Spiegelman's Q_β virus and Eigen's nucleotide soup. The objective should be once again to work from both ends, from the top down and from the bottom up, and to find out where in the middle the two ends meet. From the top, we need to find a suitable creature, an enucleated cell that has lost its replicative apparatus but still preserves the functions of metabolism and homeostasis, and we need to keep it alive artificially while stripping it gradually of inessential molecular components. We may hope in this way, with many trials and errors, to find out roughly the irreducible minimum degree of complication of a homeostatic apparatus. From the bottom, we need to experiment with synthetic populations of molecules confined in droplets in the style of Oparin, adding various combinations of catalysts and metabolites until a lasting homeostatic equilibrium is achieved. If we are lucky, we may find that the experiments from the top and those from the bottom show some degree of convergence. Insofar as they converge, they will indicate a possible pathway that life might have followed in its original progress from chaos to homeostasis.

These suggestions for future experiments probably sound naive and simpleminded to experimenters whose daily lives are spent in constant battle against the recalcitrance of real cells and real chemicals. I do not know when experiments along the lines that I have suggested will become feasible. I suggest them with diffidence, being myself incapable of doing an experiment even in my own field of physics. Nevertheless, I make these suggestions with serious intent. If I did not believe that such experiments are potentially important, I would not have ventured to talk about the origin of life in the first place. If a theoretical physicist has anything of value to say about the fundamental problems of biology, it can only be through making suggestions for new types of experiment. Half a century ago, Erwin Schrödinger suggested to biologists that they should investigate experimentally the molecular structure of the gene. That suggestion turned out to be timely. I am now suggesting that biologists investigate experimentally the population structure of homeostatic systems of molecules. If I am lucky, this suggestion may also turn out to be timely.

Before leaving the subject of future experiments, I would like to add some remarks about computer simulations. In population biology applied to animals and plants, the computer is a source of experimental data at least as important as field observation. Computer simulations of population dynamics are indispensable for the planning of field observations and for the interpretation of results. Computer simulations are not only quicker than field observations but also cheaper. Every serious program of research in population biology includes computer simulations as a matter of course. Because the origin of life is a problem in the population biology of molecules, computer simulations are essential here too. The simulations of the Oparin theory summarized by Lancet (Segré and Lancet, 1999) are a good beginning, but they still have far to go.

None of the models incorporates enough details of the chemistry to provide a realistic test of the theory.

Ursula Niesert's computer simulations of the Eigen hypercycle model of the origin of life (Niesert et al., 1981) exposed several serious weaknesses of that model. As Niesert observed as a result of her simulations, the failures of the hypercycle model are mainly due to the fact that a single RNA molecule is supposed to be performing three separate functions simultaneously. The three functions are replicating itself with the help of a replicator molecule to which it is specifically adapted, carrying a message to promote the synthesis of another molecule, and acting as agent for the specific transfer of amino acids. The computer models show that RNA molecules have a natural tendency to specialize. They prefer to perform a single function well rather than to perform three functions badly. This conclusion is not surprising. In the natural ecology of species, it is a general rule that most species survive by becoming specialists. Niesert's simulation showed that the same rule applies in the ecology of molecules in the hypercycle model. Her criticism of the model enables us to understand it better and perhaps to improve it. In the same way, computer simulations of models of the origin of homeostasis should show us what is wrong with the models and help us to replace them with better ones. Like the hypercycle model, models of homeostasis are likely to be vulnerable to the three dangers that Niesert described in her paper. As soon as realistic models of homeostatic populations become available, computer simulations will probably reveal a variety of additional catastrophes to which they are liable. Only when we have explored all the possible modes of breakdown of homeostasis will we have a right to say that we understand what homeostasis means. Computer simulations will be essential to the growth of such understanding. In our search for an answer to the question of why life is so complicated, biological and chemical experiments and computer simulations must always go hand in hand.

Computer simulations of biological evolution were begun long ago by Nils Barricelli, using the original von Neumann computer in Princeton (Barricelli, 1957; Dyson, 1997). Beginning in 1953, working within the constraints of a machine with extremely limited memory and programming directly in machine language because programming languages had not yet been invented, Barricelli successfully simulated the evolution of an ecology of numerical organisms. He observed the spontaneous origin of the phenomena of parasitism and symbiosis. He showed how Darwinian selection could lead to the evolution of complexity from simple beginnings. Unfortunately von Neumann, who had invited him to work on the computer, left Princeton in 1954 and died in 1957. It seems that von Neumann never became aware of Barricelli's achievements. Barricelli was ignored and forgotten, both by biologists and by computer scientists. His name should have been on the list of illustrious predecessors, but he never became illustrious. Thirty years later, a new generation of computer scientists with vastly greater resources began again where Barricelli stopped, performing a variety of evolutionary simulations that they called "Artificial Life." The most lifelike of the new simulations is a program called Tierra designed by Thomas Ray, a biologist who studied the ecology of plants in a real rain-forest in Costa Rica before turning his attention to simulated ecologies (Ray, 1994). The Tierra program dramatically demonstrates the phenomenon of "punctuated equilibrium" in the evolution of an artificial ecology. As the evolution runs freely on the computer, it often happens that the population structure remains in a roughly constant equilibrium state for hundreds or thousands of generations, and then a mutation causes rapid multiplication of a new species and a sudden shift of the ecology to a new equilibrium. Each time there is a shift to a new equilibrium, not only the morphology of individuals but the patterns of their behavior and mutual relations will change.

The simulated evolution experiments of Barricelli and Ray do not directly address the problem of the origin of life. They begin with a human-designed creature and explore how its progeny evolve. Ray's purpose is not to imitate organic life but to evolve a new kind of life. He is using the computer experiments as a tool to understand the nature of life in general, not to understand the nature of organic life on earth in particular. He says, "These are not models of life, but independent instances of life." The Tierra program is very far from being a realistic simulation of anything that ever lived on earth.

Nevertheless, it is more realistic than the hypercycle programs of Eigen and Niesert. It includes at least some of the complexities of a real ecology. In the future, programs like Tierra will go much further in the direction of realism and may in the end come to grips with the formidable problem of simulating the origin of metabolism.

Other Questions Suggested by the Toy Model

I now return to the toy model of Chapter 3 and examine some other questions that it raises. The questions are not specific to this particular model. They will arise for any model of the origin of life in which we have molecular populations achieving metabolism and homeostasis before they achieve replication. The questions refer not to the model itself but to the implications of the model for the subsequent course of biological evolution. I comment briefly on each question in turn. After another twenty years of progress in biological research, we may perhaps know whether my tentative answers are correct.

Were the first living creatures composed of molecules resembling proteins, or molecules resembling nucleic acids, or a mixture of the two?

I have already stated my reasons for preferring proteins. I prefer proteins partly because my model works well with ten species of monomer and works badly with four species, partly because amino acids fit the requirements of prebiotic chemistry better than nucleotides, and partly because I am attracted by the Margulis vision of parasitism as a driving force of early evolution and I like to put nucleic acids into the role of primeval parasites. None of these reasons is scientifically compelling.

At what stage did random genetic drift give way to natural selection?

The model has life originating by neutral evolution according to the ideas of Kimura (Kimura, 1970, 1983). A population confined to a cell crosses the saddle point to the ordered state by random genetic drift. The model does not allow natural selection to operate because it does not allow the populations in cells to grow or to shrink. So long as there is no multiplication and elimination of cells, there can be no natural selection. However, once a cell has reached the ordered state as defined in the model, it can go beyond the model and pass into a new phase of evolution by growing new sites for adsorption and assimilation of monomers from its environment. A cell that increases its number N of adsorption sites will quickly become stabilized against reversion to the disordered state, for the lifetime of the ordered state increases exponentially with N. It can then continue to grow until some physical disturbance causes it to divide. If it divides into two cells, there is a good chance that both daughter populations contain a sufficient assortment of catalysts to remain in the ordered state. The processes of growth and division can continue until the cells begin to exhaust the supply of nutrient monomers. When the monomers are in short supply, some cells will lose their substance and die. From that point on, evolution will be driven by natural selection.

As soon as natural selection begins to operate, there will be an enormous advantage accruing to any cell that acquires the knack of dividing itself spontaneously instead of waiting for some external process such as wave motion or turbulent flow to break it apart. At first, spontaneous division might be an accidental consequence of a tendency of the cell surface to weaken as the cell expands in volume. At a later stage, the weakening of the surface and the subsequent spontaneous division would become organized and integrated into the metabolic cycle of the cell. Cells would then be competing with one another in a straightforward Darwinian fashion with the prize of survival going to those that had learned to grow and divide most rapidly and reliably. In this way the processes of natural selection would have been well established long before the cells had acquired anything resembling the modern machinery of cell division.

Does the model contradict the Central Dogma of molecular biology?

The Central Dogma says that genetic information is carried only by nucleic acids and not by proteins. The dogma is true for all contemporary organisms with the possible exception of the prion agents responsible for scrapie and kuru. Whether or not the prion turns out to be a true exception to the dogma, my model implies that the dogma was untrue for the earliest forms of life. According to the model, the first cells passed genetic information to their off-spring in the form of catalysts that were probably molecules similar to proteins. The main requirement of the model is that the catalysts were similar to proteins in their complexity and variety. There is no logical reason why a population of molecules mutually catalyz-ing each other's synthesis should not serve as a carrier of genetic information.

The question of how much genetic information can be carried by a population of molecules without exact replication is intimately bound up with the question of the nature of homeostasis. Home-ostasis is the preservation of the chemical architecture of a popu-lation in spite of variations in local conditions and in the numbers of molecules of various kinds. Genetic information is carried in the architecture and not in the individual components. But we do not know how to define architecture or how to quantify homeosta-sis. Lacking a deep understanding of homeostasis, we can use the crude method of Chapter 3 to calculate how many items of ge-netic information the homeostatic machinery of a cell may be able to preserve. The amount of information turns out to be roughly equal to the information contained in a replicative apparatus with the same number of active components. The calculation shows that the Central Dogma is not a logical necessity. The Central Dogma is true in the modern world because of a historical accident. The accident was the invasion of primitive cells by nucleic acids. The Central Dogma need not have been true before the accident hap-pened.

It seems to be true, both in the world of cellular chemistry and in the world of ecology, that homeostatic mechanisms have a general tendency to become complicated rather than simple. Homeostasis seems to work better with an elaborate web of interlocking cycles than with a small number of cycles operating separately. Why this is so we do not know. We are back again with the question, why is life so complicated. But the prevalence of highly complex homeo-static systems, whether we understand the reasons for it or not, is a fact. This fact is additional evidence confirming our conclusion that large amounts of information are expressed in the architecture of molecular populations without nucleic acid software and without apparatus for exact replication.

How did nucleic acids originate?

We saw in Chapter 2 that nucleic acids are chemical cousins of the ATP molecule, which is the chief energy carrier in the metabolism of modern cells. I like to use this curious coincidence to explain the origin of nucleic acids as a disease arising in some primitive cell from a surfeit of ATP. The Margulis picture of evolution converts the nucleic acids from their original status as indigestible by-products of ATP metabolism to disease agents, from disease agents to parasites, from parasites to symbionts, and finally from symbionts to fully integrated organs of the cell.

How did the modern genetic apparatus evolve?

The modern genetic apparatus is enormously fine-tuned and must have evolved over a long period from simpler beginnings. Perhaps some clues to its earlier history will be found when the structure of the modern ribosome is explored and understood in detail. The following sequence of steps is a possible pathway to the modern genetic apparatus, beginning with a cell that has RNA established as a self-reproducing cellular parasite but not yet performing a genetic function for the cell: (a) nonspecific binding of RNA to free amino acids activating them for easier polymerization; (b) specific binding of RNA to catalytic sites to give them structural precision; (c) RNA bound to amino acids becomes transfer RNA; (d) RNA bound to catalytic sites becomes ribosomal RNA; (e) catalytic sites evolve from special purpose to general purpose by using transfer RNA instead of amino acids for recognition; (f) recognition unit splits off from ribosomal RNA and becomes messenger RNA; and (g) ribosomal structure becomes unique as the genetic code takes over the function of recognition. This is only one of many possible pathways that might have led to the evolution of the genetic code. The essential point is that all such pathways appear to be long and tortuous. In my opinion, the metabolic machinery of proteins and the parasitic self-replication of nucleic acids must have been in place before the evolution of the elaborate translation apparatus linking the two systems could begin.

How late was the latest common ancestor of all living species?

The universality of the genetic code shows that the latest common ancestor of all living creatures already possessed a complete genetic apparatus of the modern type. The geological record tells us that cells existed very early, as long as 3.5 eons ago. It is generally assumed that the earliest cells that are preserved as microfossils already possessed a modern genetic apparatus, but this assumption is not based on concrete evidence. It is possible that the evolution of the modern genetic apparatus took eons to complete. The ancient microfossils may date from a time before there were genes and ribosomes. The pace of evolution may have accelerated after the genetic code was established, allowing the development from ancestral procaryote to eucaryotic cells and multicellular organisms to be completed in less time than it took to go from primitive cell to ancestral procaryote. It is therefore possible that the latest common ancestor came late in the history of life, perhaps as late as half-way from the beginning.

Does there exist a chemical realization of my model, for example, a population of a few thousand amino acids forming an association of polypeptides that can catalyze each other's synthesis with 80-percent accuracy? Can such an association of molecules be confined in a droplet and supplied with energy and raw materials in such a way as to maintain itself in a stable homeostatic equilibrium? Does the addition of a solid surface, such as a clay crystal or a metal sulphide membrane, help to stabilize the equilibrium?

These are the crucial questions that only experiment can answer.

What will happen to my little toy model when the problem of the origin of life is finally solved?

This is the last question raised by the model and it is easily answered. The answer was given nearly two hundred years ago by my favorite poet, William Blake ("A Vision of the Last Judgment," Rossetti MS, 1810):

To be an Error and to be Cast out is a part of God's design.

WIDER IMPLICATIONS

At the end of his book *What is Life?*, Schrödinger put a four-page epilogue with the title "On determinism and free will." He there states his personal philosophical viewpoint, his reconciliation between his objective understanding of the physical machinery of life and his subjective experience of free will. He writes with a clarity and economy of language that have rarely been equaled. I will not try to compete with Schrödinger in summing up in four pages the fruits of a lifetime of philosophical reflection. Instead I will use my last pages to discuss some of the wider implications of our thoughts about the origin of life, not for personal philosophy but for other areas of science. I use the word science here in a broad sense, including social as well as natural sciences. The sciences that I have particularly in mind are ecology, economics, and cultural history. In all these areas we are confronting the same question that is at the root of the problem of understanding the origin of life: Why is life so complicated? It may be that each of these areas has something to learn from the others.

The concept of homeostasis can be transferred without difficulty from a molecular context to ecological, economic, and cultural contexts. In each area we have the unexplained fact that complicated homeostatic mechanisms are more prevalent and seem to be more effective than simple ones. This is most spectacularly true in the domain of ecology, where a typical stable community, for example a few acres of woodland or a few square feet of grassland, comprises thousands of diverse species with highly specialized and interdependent functions. But a similar phenomenon is visible in economic life and in cultural evolution. The open market economy and the culturally open society, notwithstanding all their failures and deficiencies, seem to possess a robustness that centrally planned economies and culturally closed societies lack. The homeostasis provided by unified five-year economic plans and by unified political control of culture does not lead to a greater stability of economies and cultures. On the contrary, the simple homeostatic mechanisms of central control have generally proved more brittle and less able to cope with historical shocks than the complex homeostatic mechanisms of the open market and the uncensored press.

But I did not intend this book to be a political manifesto in defense of free enterprise. My purpose in mentioning the analogies between cellular and social homeostasis was not to draw a political moral from biology but rather to draw a biological moral from ecology and social history. Fortunately, I can claim the highest scientific authority for drawing the moral in this direction. It is well known to historians of science that Charles Darwin was strongly influenced in his working out of the theory of evolution by his readings of the political economists from Adam Smith to Malthus and McCullough. Darwin himself said of the theory: "This is the doctrine of Malthus applied to the whole animal and vegetable kingdom." What I am proposing is to apply in the same spirit the doctrines of modern ecology to the molecular processes within a primitive cell. In our present state of ignorance we have a choice between two contrasting images to represent our view of the possible structure of a creature newly emerged at the first threshold of life. One image is the hypercycle model of Eigen, with molecular structure tightly linked and centrally controlled, replicating itself with considerable precision and achieving homeostasis by strict adherence to a rigid pattern. The other image is the "tangled bank" of Darwin, an image that Darwin put at the end of his *Origin of Species* to make vivid his answer to the question, What is life?, an image of grasses and flowers and bees and butterflies growing in tangled profusion without any discernible pattern, achieving homeostasis by means of a web of interdependences too complicated for us to unravel. The tangled bank is the image that I have in mind when I try to imagine what a primeval cell would look like. I imagine a collection of molecular species that are tangled and interlocking like the plants and insects in Darwin's microcosm. This was the image that led me to think of error tolerance as the primary requirement for a model of a molecular population taking its first faltering steps toward life. Error tolerance is the hallmark of natural ecological communities, of free market economies, and of open societies. I believe it must have been a primary quality of life from the very beginning. But replication and error tolerance are naturally antagonistic principles. That is why I like to exclude replication from the beginnings of life, to imagine the first cells as error-tolerant tangles of nonreplicating molecules, and to introduce replication as an alien parasitic intrusion at a later stage. Only after the alien intruder has been tamed is the reconciliation between replication and error tolerance achieved in a higher synthesis through the evolution of the genetic code and the modern apparatus of ribosomes and chromosomes.

The modern synthesis reconciles replication with error tolerance by establishing the division of labor between hardware and software, between the genetic apparatus and the gene. In the modern cell, the hardware of the genetic apparatus is rigidly controlled and error intolerant. The hardware must be error intolerant to maintain the accuracy of replication. But the error tolerance that I like to believe was inherent in life from its earliest beginnings has not been lost. The burden of error tolerance has merely been transferred to the software. In the modern cell, with the infrastructure of hardware firmly in place and subject to a strict regime of quality control, the software is free to wander, to make mistakes, and occasionally to be creative. The transfer of architectural design from hardware to software allowed the molecular architects to work with a freedom and creativity that their ancestors before the transfer could never have approached. A similar transfer of architectural design from embryo to adult probably caused the outburst of evolutionary novelty that we call the Cambrian explosion.

The analogies between the genetic evolution of biological species and the cultural evolution of human societies have been brilliantly explored by Richard Dawkins in his book *The Selfish Gene* (Dawkins, 1976). The book is mainly concerned with biological evolution. The cultural analogies are pursued only in the last chapter. Dawkins's main theme is the tyranny that the rigid demands of the replication apparatus have imposed upon all biological species throughout evolutionary history. Every species is the prisoner of its genes and is compelled to develop and to behave in such a way as to maximize their chances of survival. Only the genes are free to experiment with new patterns of behavior. Individual organisms must do what their genes dictate. This tyranny of the genes has lasted for 3 eons and has been precariously overthrown only in the last hundred thousand years by a single species, Homo sapiens. We have overthrown the tyranny by inventing symbolic language and culture. Our behavior patterns are now to a great extent culturally rather than genetically determined. We can choose to keep a defective gene in circulation because our culture tells us not to let hemophiliac children die. We have stolen back from our genes the freedom to make choices and to make mistakes.

In his last chapter Dawkins describes a new tyrant that has arisen within human culture to take the place of the old. The new tyrant is the "meme," the cultural analogue of the gene. A meme is a behavioral pattern that replicates itself by cultural transfer from individual to individual instead of by biological inheritance. Examples of memes are religious beliefs, linguistic idioms, fashions in art and science and in food and clothes. Almost all the phenomena of evolutionary genetics and speciation have their analogues in cultural history, with the meme taking over the functions of the gene. The meme is a self-replicating unit of behavior like the gene. The meme and the gene are equally selfish. The history of human culture shows us to be as subject to the tyranny of our memes as other species are to the tyranny of genes. But Dawkins ends his discussion with a call for liberation. Our capacity for foresight gives us the power to transcend our memes just as our culture gave us the power to transcend our genes. We, he says, alone on earth, can rebel against the tyranny of the selfish replicators.

Dawkins's vision of the human situation as a Promethean struggle against the tyranny of the replicators contains important elements of truth. We are indeed rebels by nature, and his vision explains many aspects of our culture that would otherwise be mysterious. But his account leaves out half the story. He describes the history of life as the history of replication. Like Eigen, he believes that the beginning of life was a self-replicating molecule. Throughout his history, the replicators are in control. In the beginning, he says, was simplicity. The point of view that I am expounding in these lectures is precisely the opposite. In the beginning, I am saying, was complexity. The essence of life from the beginning was homeostasis based on a complicated web of molecular structures. Life by its very nature is resistant to simplification, whether on the level of single cells or ecological systems or human societies. Life could tolerate a precisely replicating molecular apparatus only by incorporating it into a translation system that allowed the complexity of the molecular web to be expressed in the form of software. After the transfer of complication from hardware to software, life continued to be a complicated interlocking web in which the replicators were only one component. The replicators were never as firmly in control as Dawkins imagined. In my version the history of life is counterpoint music, a two-part invention with two voices, the voice of the replicators attempting to impose their selfish purposes upon the whole network and the voice of homeostasis tending to maximize diversity of structure and flexibility of function. The tyranny of the replicators was always mitigated by the more ancient cooperative structure of homeostasis that was inherent in every organism. The rule of the genes was like the government of the old Hapsburg Empire: *Despotismus gemildert durch Schlamperei,* or "despotism tempered by sloppiness."

As the grandfather of a pair of five-year-old identical twins, I see every day the power of the genes and the limits to that power. George and Donald are physically so alike that in the bathtub I cannot tell them apart. They not only have the same genes but have shared the same environment since the day they were born. And yet, they have different brains and are different people. Life has escaped the tyranny of the genes by evolving brains with neural connections that are not genetically determined. The detailed structure of the brain is partly shaped by genes and environment and is partly random. Earlier, when the twins were two years old, I asked their older brother how he tells them apart. He said, "Oh, that's easy. The one that bites is George." Now that they are five years old, George is the one who runs to give me a hug, and Donald is the one who keeps his distance. The randomness of the synapses in their brains is the creative principle that makes George George and Donald Donald.

One of the most interesting developments in modern genetics is the discovery of "junk DNA," a substantial component of our cellular inheritance that appears to have no biological function. Junk DNA is DNA that does us no good and no harm, merely taking a free ride in our cells and taking advantage of our efficient replicative apparatus. The prevalence of junk DNA is a striking example of the sloppiness that life has always embodied in one form or another. It is easy to find in human culture the analogue of junk DNA. Junk culture is replicated together with memes, just as junk DNA is replicated together with genes. Junk culture is the rubbish of civilization: television commercials, Internet spam, astrology, and political propaganda. Tolerance of junk is one of life's most essential characteristics. I would be surprised if the first living cell had not been at least 25 percent junk.

In every sphere of life, whether cultural, economic, ecological, or cellular, the systems that survive best are those that are not too fine-tuned to carry a large load of junk. And so, I believe, it must have been at the beginning. The early evolution of life probably followed the same pattern as the development of the individual human brain, beginning with a huge assortment of random connections and slowly weeding out by trial and error the connections that made no sense. George and Donald are different people because they started life with different random samples of neurological junk in their heads. The weeding out of the junk is never complete. Adult humans are only a little more rational than five-year-olds. Too much weeding destroys the soul.

That is the end of my story, and it brings me back to the beginning. I have been trying to imagine a framework for the origin of life, guided by a personal philosophy that considers the primal characteristics of life to be homeostasis rather than replication, diversity rather than uniformity, the flexibility of the genome rather than the tyranny of the gene, the error tolerance of the whole rather than the precision of the parts. The framework that I have found is an abstract mathematical model that is far too simple to be true. But the model incorporates in a crude fashion the qualitative features of life that I consider essential: looseness of structure and tolerance of errors. The model fits into an overall view of life and evolution that is more relaxed than the traditional view. The new and looser picture of evolution is strongly supported by recent experimental discoveries in the molecular biology of eucaryotic cells. Edward Wilson, who was also my illustrious predecessor as Tarner Lecturer in Cambridge (Wilson, 1982), describes the new picture of the eucaryotic genome as "a rainforest with many niches occupied by a whole range of elements, all parts of which are in a dynamic state of change." My philosophical bias leads me to believe that Wilson's picture describes not only the eucaryotic genome but the evolution of life all the way back to the beginning. I hold the creativity of quasi-random complicated structures to be a more important driving force of evolution than the Darwinian competition of replicating monads. But philosophy is nothing but empty words if it is not capable of being tested by experiment. If my remarks have any value, it is only insofar as they suggest new experiments. I leave it now to the experimenters to see whether they can condense some solid facts out of this philosophical hot air.

Bibliography

Anderson, P. W. (1983). Suggested model for prebiotic evolution: The use of chaos. *Proc. Nat. Acad. Sci. USA*, **80**, 3386–3390.

Barricelli, N. A. (1957). Symbiogenetic evolution processes realized by artificial methods. *Methodos*, **9**, 143–182.

Cairns-Smith, A. G. (1982). *Genetic Takeover and the Mineral Origins of Life*. New York: Cambridge University Press.

Cairns-Smith, A. G. (1985). *Seven Clues to the Origin of Life*. Cambridge: Cambridge University Press.

Cech, T. R. (1993). The efficiency and versatility of catalytic RNA: Implications for an RNA world. *Gene*, **135**, 33–36.

Cech, T. R. and B. L. Bass (1986). Biological catalysis by RNA. *Ann. Rev. Biochem.* **55**, 599–629.

Chyba, C. F. and G. D. McDonald (1995). The origin of life in the solar system: Current issues, *Ann. Rev. Earth Planet. Sci.*, **23**, 215–249.

Davidson, E. H., K. J. Peterson, and R. A. Cameron (1995). Origin of bilaterian body plans: Evolution of developmental regulatory mechanisms, *Science*, **270**, 1319–1325.

Davies, P. (1998). *The Fifth Miracle: The Search for the Origin of Life*. London: Penguin Books.

Dawkins, R. (1976). *The Selfish Gene*. New York: Oxford University Press.

Dyson, F. J. (1982). A model for the origin of life. *J. Mol. Evol.*, **18**, 344–350.

Dyson, G. B. (1997). *Darwin Among the Machines*. New York: Addison-Wesley. See chapter 7 for Merezhkovsky, Barricelli, and Ray.

Eigen, M., W. Gardiner, P. Schuster, and R. Winckler-Oswatitch (1981). The origin of genetic information. *Sci. Am.*, **244**(4), 88–118.

Gajdusek, D. C. (1977). Unconventional viruses and the origin and disappearance of kuru. *Science*, **197**, 943–960.

Gilbert, W. (1986). The RNA world. *Nature*, **319**, 618.

Gold, T. (1992). The deep, hot biosphere. *Proc. Nat. Acad. Sci. USA*, **89**, 6045–6049.

Gold, T. (1998). *The Deep Hot Biosphere*, New York: Springer-Verlag.

Huber, C. and G. Wächtershäuser (1998). Peptides by activation of amino acids with CO on (Ni, Fe)S surfaces: Implications for the origin of life. *Science*, **281**, 670–672.

Joyce, G. F. (1989). RNA evolution and the origins of life. *Nature*, **338**, 217–224.

Jukes, T. (1997). Oparin and Lysenko. *J. Mol. Evol.*, **45**, 339–341.

Khakhina, L. N. (1992). *Concepts of Symbiogenesis: A Historical and Critical Study of the Research of Russian Botanists*, trans. L. Merkel, ed. L. Margulis and M. McMenamin. New Haven: Yale University Press.

Kimura, M. (1970). Stochastic processes in population genetics. In *Mathematical Topics in Population Genetics*, ed. K. I. Kojima, pp. 178–209. Berlin: Springer-Verlag.

Kimura, M. (1983). *The Neutral Theory of Molecular Evolution*. New York: Cambridge University Press.

Margulis, L. (1970). *Origin of Eucaryotic Cells*. New Haven: Yale University Press.

Margulis, L. (1981). *Symbiosis in Cell Evolution*. San Francisco: Freeman and Co.

Margulis, L. and D. Sagan (1995). *What is Life?* New York: Simon and Schuster.

Merezhkovsky, K. S. (1909). *Theory of Two Plasms as the Basis of Symbiogenesis: A New Study on the Origin of Organisms*, in Russian. Kazan: Publishing office of the Imperial Kazan University.

Miller, S. M. and L. E. Orgel (1974). *The Origins of Life on the Earth*. Englewood Cliffs, NJ: Prentice-Hall, Inc.

Mojzsis, S. J., G. Arrhenius, K. D. McKeegan, T. M. Harrison, A. P. Nutman, and C. R. L. Friend (1996). Evidence for life on Earth before 3800 million years ago, *Nature*, **384**, 55–59.

Niesert, U., D. Harnasch, and C. Bresch (1981). Origin of life between Scylla and Charybdis. *J. Mol. Evol.*, **17**, 348–353.

Nisbet, E. G. (1995). Archean ecology: A review of evidence for the early development of bacterial biomes, and speculations on the development of a global-scale biosphere. In *Early Precambrian Processes*, ed. M. P. Coward and A. C. Ries, Geological Society Special Publications, No. 95, pp. 27–51.

Oparin, A. I. (1957). *The Origin of Life on the Earth*, 3rd ed., trans. Ann Synge. Edinburgh: Oliver and Boyd.

Perutz, M. F. (1989). *Erwin Schrödinger's "What is Life?" and Molecular Biology.* In M. F. Perutz, *Is Science Necessary?* New York: Oxford University Press, pp. 234–251.

Prusiner, S. B. (1982). Novel proteinaceous infectious particles cause scrapie. *Science*, **216**, 136–144.

Prusiner, S. B. (1991). Molecular biology of prion diseases. *Science*, **252**, 1515–1522.

Ray, T. S. (1994). An evolutionary approach to synthetic biology: Zen and the art of creating life. *Artificial Life*, **1**, 179–209.

Russell, M. J., R. M. Daniel, A. J. Hall, and J. Sherringham (1994). A hydrothermally precipitated catalytic iron sulphide membrane as a first step toward life. *J. Mol. Evol.*, **39**, 231–243.

Santoro, S. W. and G. F. Joyce (1997). A general purpose RNA-cleaving DNA enzyme. *Proc. Nat. Acad. Sci. USA*, **94**, 4262–4266.

Schrödinger, E. (1944). *What is Life? The Physical Aspect of the Living Cell.* Cambridge: Cambridge University Press.

Segré, D. and D. Lancet (1999). A statistical chemistry approach to the origin of life. *Chemtracts—Biochem. Mol. Biol.*, **12**(6), 382–397.

Spiegelman, S. (1967). An in vitro analysis of a replicating molecule. *Am. Sci.*, **55**, 3–68.

Timoféeff-Ressovsky, N. W., K. G. Zimmer, and M. Delbrück (1935). Ueber die Natur der Genmutation und der Genstruktur. *Nachr. Ges. Wiss. Göttingen*, **6** NF(13), 190–245.

Von Neumann, J. (1948). The General and Logical Theory of Automata. Lecture given 1948. In *Cerebral Mechanisms in Behavior—The Hixon Symposium*, ed. L. A. Jeffress, pp. 1–41, New York: John Wiley, 1951; and in *J. von Neumann, Collected Works*, Vol. 5, ed. A. H. Taub, pp. 288–328, New York: MacMillan, 1961–63.

Wächtershäuser, G. (1992). Groundworks for an evolutionary biochemistry: The iron-sulphur world. *Prog. Biophys. Mol. Biol.*, **58**, 85–201.

Wächtershäuser, G. (1997). The origin of life and its methodological challenge. *J. Theoret. Biol.*, **187**, 483–494.

Wilson, E. O. (1982). Remarks quoted by R. Lewin. *Science*, **216**, 1091–1092.

Wright, C. W. and G. F. Joyce (1997). Continuous in vitro evolution of catalytic function. *Science*, **276**, 614–617.

Index